DODGING THE DEATH RAYS

A Medical Look at Our Deep Space Policy

Alvin L. Ureles M.D.

Professor of Medicine Emeritus
University of Rochester School of Medicine

AuthorHouse™
1663 Liberty Drive
Bloomington, IN 47403
www.authorhouse.com
Phone: 1 (800) 839-8640

© 2015 Alvin L. Ureles M.D. All rights reserved.

No part of this book may be reproduced, stored in a retrieval system, or transmitted by any means without the written permission of the author.

Published by AuthorHouse   06/02/2015

ISBN: 978-1-5049-1281-5 (sc)
ISBN: 978-1-5049-1282-2 (e)

Library of Congress Control Number: 2015908156

Print information available on the last page.

Any people depicted in stock imagery provided by Thinkstock are models, and such images are being used for illustrative purposes only. Certain stock imagery © Thinkstock.

This book is printed on acid-free paper.

Because of the dynamic nature of the Internet, any web addresses or links contained in this book may have changed since publication and may no longer be valid. The views expressed in this work are solely those of the author and do not necessarily reflect the views of the publisher, and the publisher hereby disclaims any responsibility for them.

This book is dedicated to Dr. Saul Hertz, mentor and scholar, who trained me to see the bright road and the dark path of ionizing radiation

CONTENTS

Introduction ix
1   Sunshine 101 1
2   Ionizing Radiation Everywhere 11
3   The Death Rays 21
4   The Villain From Outer Space 31
5   Plans To Return To The Moon 38
6   The Long March to Mars 49
7   The Ultimate NASA Mission 58
8   Understanding Space Bio-hazards 66
9   Why We Dodge 78
10 Defense 106
11 Final Thoughts 117
References 127
Bibliography 135
Appendix A. Space Radiobiology 137
Appendix B. Timeline For Cosmic Rays 143
Appendix C. Common Health Problems In Space Flight That Might Potentiate The Radiation Hazzards 149
Appendix D. Quantum Comments 153

INTRODUCTION

The intent of this book is to help you understand the radiation hazards of deep space flight since you will soon be asked to support a national space enterprise with "astronomical" costs. Understanding leads an intelligent citizenry to seek answers and ask the right questions.

I have tried to present the astrophysical and biomedical material in a manner that is clear and ready to grasp and have placed *"gray material"* throughout the text for those desiring more details along with an Appendix.

Now after decades of teaching and practicing medicine I reflect on what has prompted me to want to write about this issue.

At the end of WWII I returned from service to be appointed as a Harvard Research Fellow to work with Dr. Saul Hertz who had pioneered the experimental applications of radioactive iodine and was returning from service in the Navy ready to develop the new and exciting field of Nuclear Medicine. It was for me the beginning of a lifetime involvement in the application of ionizing radiation to save lives.

All of this ultimately led to my returning home to Rochester, N.Y. with an appointment at the University of Rochester Medical Center's Department of Medicine and as an associate in Radiology where I joined with Dr. Philip Rubin, the eventual Grand Master of Radiotherapy and Radiation Biology. I certified as a specialist in Nuclear Medicine as well as Internal Medicine, joined both Societies and took on a heavy teaching schedule and cared for patients.

The story could well have ended there but life is full of surprises and enhancements.

Stepping down in the eighties from my appointment as Chief of Medicine at one of our teaching hospitals the staff kindly prepared a lectureship in my name and knowing my lifelong hobby and fascination with Astronomy had contacted NASA, inviting an "active in line" astronaut to give the first Ureles Lecture.

The presentation was an inspiring success. It offered me insight as to what these young heroes undergo in both physical and mental training. It opened up for me a fresh look at the medical aspects of space flight, a probe into flight physiology and a new look at our space program.

The following year another "active in line" astronaut gave a second stellar Ureles Lecture. The die was cast. I was a NASA fan. These two young astronauts fostered in me an evolving world of study with special regard for the health and welfare of all the many remarkable young men and women that serve in our Space Program. I began lecturing on The Health Concerns in Human Space Flight and was pleased to see my material incorporated into one of the astronomy courses of the college.

So here we are in the early 21st Century with big space plans knocking at the door of Congress and many gifted people with loud voices and good intentions arguing for and against the American Dream of putting humans back on the Moon, settling humans on Mars and the Moon and certainly perhaps the most critical decision of all some role for astronauts in the methodology needed for deflecting those dangerous planet-destroying earth crossing objects that could be heading our way.

There is an understandable concern for the dollar cost for these enterprises by a nation struggling with debt, cities wracked with poverty, roads and bridges long neglected, burgeoning needs for areas hit by drought, wild fire, hurricanes, tornados and earthquakes.

But there is also the innate curiosity of man, the restless need to explore the unknown, the flaming desire to be the first, unwavering national pride and the gold at the end of the rainbow.

This book aims to address one aspect of the human cost of 21st Century Space Initiatives. Are we ready to put a man and woman in deep space over time and create extra-planetary habitats? If not now, when? And with what caveats and containments?

The stumbling block to be addressed is the very same physical phenomena to which I had devoted a good part of my life, namely the radiobiology of ionizing radiation. But this is the ionizing radiation of space, particularly deep outer space and this is a far cry from the radiation of the medical discipline that saves lives.

With a careful look at the data the hope is to bring some conclusions to the table on this important subject. Above all the intent is to explain the accumulated evidence in terms of the risks and available remedies in levels of detail so that every citizen has an opportunity to pick and choose how much they wish to understand and how they want to vote.

Chapter One

SUNSHINE 101

First, a quick refresher for those who have been away from science for a spell.

This book will be full of references to "RAYS", and it is important to define what is meant here by the term since there is some wiggle room about the definition among the scientific community. There will be no math or complex physics and chemistry; plain noble concepts only so take-home points are really taken home.

The Rays under discussions are both

1. Energetic massless particles that make up the broad electromagnetic (EM) spectrum, which includes the small narrow band which we call "light". We will be concerned with special EM rays, the X-Rays and Gamma Rays. These rays behave both like waves and minute packets of energy called "quanta". Don't try to visualize this behavior. You can't. It is just your

introduction to the "Alice in Wonderland" aspect of modern quantum physics.

2. Subatomic particles containing mass, great energy and (except for rare exceptions) electrical charge. They emanate from some active astronomical source such as our Sun or some stellar catastrophe somewhere out in our Galaxy. They move at great speed and I shall have much to say about them.

Recall that all mater is made up of atoms and that all of nature is made up of 90 different atoms. Atoms are very small, 10^{10th} meters in size, a stretch for viewing by even our best transmission electron microscopes. Interestingly, all natural atoms on the periodic table in spite of their increasing complexity are roughly the same size.

Hydrogen, the simplest and most common substance in the universe, is element number one. It contains a nucleus bearing one positively charged particle mass called the *proton* that holds onto an oppositely charged extremely light "orbiting mass"- the *electron*.

Think of a lead sized golf ball down in the center of the field of the Super Dome whose electrical attraction can keep a single poppy seed whizzing about the outer bleachers and across the field in three dimensions.

Yes, atoms are mostly empty space held together by strong electrical attractions with *electrons* that travel in fixed tiers with different energies that can be knocked out of their "orbits" by an intruder. Every time we add a single *proton* to an atomic nucleus we add an additional unit of mass with a positive charge and create a unique new element which must hold onto one more *electron* in order to stay neutral. So we list elements on the periodic table by their *proton* count, one addition after another.

If we add to the nucleus an electrically neutral mass about equal to the *proton* - a *neutron*, we add to the mass of the atom but we have not changed the total charge and therefore not the element. We have only created an *isotope* of that element.

Finally, if we knock off an *electron* from an element we lose the electrical balance. We have more positive *protons* than negative *electrons*. We have a simple *ion*.

If we knock all the *electrons* off a heavier element we now have a *maximized ion* whose total positive charge is equal to the *proton* count of the element. This is a "big charge".

(Quantum mechanics at a deeper level – see Appendix D - would fine tune much of this picture, but have no fear this blueprint presented will take us where we want to go.)

We shall need to focus on Ionizing Radiation (IR) which refers to the ability of certain energetic rays to do exactly what we have just described, namely, on impact with matter, such as your laptop or your cerebral cortex, knock negatively charged electrons out of their "orbits" of surrounding atoms so that the "denuded" elements and molecules are no longer electrically balanced, creating *ions*, a beehive of toxic destructive activity that break vital chemical bonds.

....

Astronomers have a particular respect for this planet on which we live. It has to do with a number of extraordinary happenstances which can be viewed on a broad pallet of responses extending from sheer luck to a spiritual gift.

Number one on this list is our planet's precise position within the solar system - the "F" words in Astronomy. Any closer to our Sun, we Fry; any further out, we Freeze and where we are in space Feels Fine.

Lucky us. Solar system gravity and our rotation holds us in this select habitable position where we are bathed in sunshine.

Sunshine comes to us from "our star," the Sun - 93 million miles away, a little over an eight minute journey to get to us at the speed of light (186 thousand miles per second!). And it has been doing this for over 4.6 billion years and will continue for another 5 billion years.

Every young student gets to understand what we adults too soon fail to appreciate as we water our gardens. Sunlight stimulates photosynthesis in the leaves of the plant kingdom. Carbon dioxide and water are absorbed, sugar is formed and vital oxygen comes out and without it all useful life vanishes

It is sunlight that evaporates the oceans and brings us rain and water to survive, converts the sterols in our skin to Vitamin D so that we have bones and are not crawling around like lumps of jelly. The sunshine warms us warm-blooded animals and has a positive influence on our mood and behavior, gives us our life, seasons, and dominates the ecosystem.

When Isaac Newton ran sunlight through a slit in his window shade and passed it through a glass prism he brought our whole naïve understanding of sunlight to a new level.

It was no more a ray of sunshine but an ordered continuum of colors, a revelation that over time we came to understand as the electromagnetic (EM) spectrum, a collection of wave/non-wave photons, progressive energies and frequencies ranging from big looping friendly slow radio waves at one end to high energy rapid short dangerous X-rays and Gamma Rays at the other end.

And to add to our amazement, all of these substituent rays travel in a vacuum at the same speed, 186,000 miles per second (nothing ever to exceed them).

In our exuberance over this discovery we note there is an embarrassingly small window of radiation on this broad EM spectrum that physicists call "white light" and we call "day light". It is this narrow band we humans use to see with our own eyes the three dimensional beauty of this wonderful world.

There are some notable facts about the sunshine's origin that play a major role in some of the fundamental problems to be dealt with in this book. We take the Sun so for granted that the busy educated mind often has a serious information gap about our star's awesome structure and function.

Yes, it is a very big mass of stellar gas (mostly hydrogen 74% with a significant amount of helium 24% and a small amount of carbon, oxygen, nitrogen and heavier atoms). A mind experiment would allow you to pack a million Earths into its bigness, or one hundred Earths lined up against its equator. But the really big story is at its core where the gasses are so dense and compressed by gravity that the extreme temperature (about 27 million degrees F.) and immense pressure converts the hydrogen gas into helium with a sliver of left over mass that is converted into a prodigious explosion of energy (*nuclear fusion*) according to Einstein's famous law (energy equals mass multiplied by the speed of light squared).

The atoms of gas in the Sun are so exceedingly hot that the electrons and protons part company and result in a chaotic charged soup called *plasma* that interacts with the electromagnetic fields, generating electric and thermal currents with vast magnetic bands and convective cells – *the granulations.*

Note once again "lucky us". There is just the right balance between gravity compressing in and the exploding core expanding out to keep our Sun in a steady state.

It is this fusion energy that makes it all happen, "burning" hydrogen, (equal to four trillion trillion 100 watt light bulbs every second!) and starting out this EM energy in the form of high frequency Gamma Radiation on a convoluted random walk from its center to the decreased temperature of the "apparent surface" that we can observe best through a telescopic filter (to protect our eyes from its blinding radiance), the so called *photosphere* where the peak EM frequency has now changed to white light to which human vision has adapted through eons of evolution.

Note that if we observe the Sun using a hydrogen alpha filter we discover a layer beyond the *photosphere* that has shifted to the red band - the *chromosphere*. It has a different depth, increased temperature and behavior. Similarly if we observe the Sun with a special telescope that blocks out natural light rays or observe a solar eclipse we discover an amazing layer of plasma beyond the chromosphere extending out into the depths of space, mysterious even today because instead of losing heat as we would expect it is a blazing fury of heat and a focus of much solar activity - the *corona and the solar wind.*

All of this is a long process for the EM rays, proceeding from the core out through a radiative zone into a convection zone, bumping into dense constituent particles and gradually losing energy. So by the time the high frequency Gamma Rays of sunshine reach the photosphere (estimates are up to a million years) much of their energies have been converted and dominated by visible light rays, but with significant amounts of Infrared, Ultraviolet, Radio and X-rays all ready for release out in every direction, including the 8.3 minute voyage to mother Earth.

We already have the technology to gather its prodigious energy. The rays that are beaming off the Sun's giant sphere are only partially interrupted by our relatively tiny Earth, but the Solar Constant (1400 Watts per square meter of earth surface) if judiciously collected on solar panels of the right size, sensitivity and price can all but eliminate the fossil fuel requirements of the world. This is radiation at its best.

Sunshine - beautiful to behold on Earth, with all the advantages to which we have referred - unfortunately has a flipside called trouble.

There are, of course, the environmental negatives from too much or too little sunshine. The ever present need to protect our ozone from depletion lest we have a consequent ultraviolet scourge; and the ever present need to modulate our greenhouse gases.

So too the personal caveats:

Do not look directly into the Sun especially through binoculars or a simple child's telescope. The retinal damage from concentrated Infra-red Rays is sustained in a flash and can cause irreversible blindness.

Do not focus sunshine with a magnifying glass. Focused Infra-red Radiation is an invitation for a blazing fire.

Do not overexpose your skin to solar radiation. Need we warn you about the skin burns, cancer and cataracts as dangers from over exposure to ultraviolet?

These are real problems over which we have some control. But, we need to examine the unpredictable dangerous and destructive aspects of solar behavior over which we have no control. We need to address them in detail since they make up our first major encounter with ionizing radiation.

Chapter Two

IONIZING RADIATION EVERYWHERE

As noted, looking at the EM spectrum there is a significant change in energy and character as we pass by visible light and Ultraviolet Rays and advance to X-Rays and Gamma Rays. These two high frequency rays, with short wave lengths and energetic beams of photons unlike their neighbors to the left, harbor so much energy they are the rays that can penetrate deeply into solid objects and for our concerns easily penetrate human tissue. These are the rays we can use to see if a bone is broken or the patient has pneumonia, or requires a tumor to be excised.

But the hallmark of this penetration is not their depth but the fact that while they are on their invasive path they are smashing into atoms of organic and inorganic molecules, breaking their links, driving off their outer orbiting electrons and forming clusters of positive and negative ions which in turn can bring to a halt the tightly scheduled complex biochemistry of life. It is from this highly destructive uncompromising process that we derive the name Ionizing Radiation (IR).

These are bad guys but are not the only bad guys. They come at us from the Sun and also from outer space, and we'll stick with the former for the moment.

Certainly we can ask the question: "If dangerous X-Ray and Gamma radiation is an intrinsic part of ordinary sunshine, how do we account for the fact that plant and animal life has survived on planet Earth?"

The answer comes from one of the most prosaic and unappreciated phenomena in our ecosystem.

We have what the Moon and Asteroids do not have and what Mars barely has - we have an *atmosphere*.

This gravity held blanket of gas (78% nitrogen, 20% oxygen, and a small amount of carbon dioxide, argon and water vapor) envelops our earth, its surface density thinning as it extends some 600 miles out. It traps the right amount of the Sun's heat and holds its oxygen in place at the lower levels so we can breathe and has the remarkable ability via the special properties of nitrogen and oxygen to absorb the dangerous right side of the spectrum past visible light- namely soaking up most of the life threatening X-Rays and Gamma Rays. (We are not even including the important O3 ozone layer that screens out the dangerous Ultraviolet B radiation.) True, the atmosphere does not function at the level of one hundred per cent but well enough to allow us to steadily evolve from the primordial ooze.

Once again, lucky us. You can be a planet in the right place, with the right amount of sunshine but if you don't have an atmospheric blanket of quality that will stay in place and selectively shield you, it is taps at sundown.

Still we are not a species that gets completely off the hook nor does the rest of the globe's plant and animal life. There is about us a constant haze of natural Earth-bound radioactivity and Cosmic Radiation that has been part of the Earth environment since the birth of the solar system. All living things have had to adapt to it. The radioactivity is characterized by ionizing radiation emanating from unstable elements that keep decaying until they form stable ones.

This radioactivity came originally from some large stars that ended their life in a total blowup– the super nova explosions that seeded interstellar space with high mass unstable elements.

(We are still getting hit from outer space by big wallops of Cosmic Rays from supernovas. They splash into the atmosphere and cause a steady rain of secondary rays that are whistling right through us right now. We shall address these shortly.)

All atoms with more than 84 protons are unstable and decay to lower elements peppering us with variable amounts of energetic ionizing photons (Gamma and X-Rays), charged particles (electrons, alpha and beta rays), twenty four hours a day through-out our lifetime with an impact that is subtle, variable, often hidden and difficult to quantify and are for the most part can be completely ignored since at their relative low dose are well tolerated by humans and most animal and plant life.

This natural background earth radiation is principally found in rocks and soil. *Radon gas* (a breakdown product of naturally occurring uranium) accounts for the majority of public exposure to ionizing radiation.

Radon is one of the rare elements in background radiation that can be dangerous when inhaled in high concentrations, as in mining. There are public health warnings even for the radon dose concentrations that sneak unobtrusively into our house cellars. The U.S. Environmental Protection Agency recommends that all house be tested for radon.

We have natural internal radiation within our bodies such as Carbon-14 and Potassium-40 passing in and out through the food chain. We will not stop eating bananas even though they harbor inescapable minute amounts of this radioactivity.

Of course we have added to the natural background radiation the radiation burden that is man-made; from nuclear production plants, both contained and leaking, nuclear waste, storage, mining and products that have radioactive components, extensive laboratory use of radioisotopes, but principally medical exposure via an increasing array of diagnostic and therapeutic tools to which every man woman and child somewhere somehow at some time will be significantly exposed. A good approximation released by The World Nuclear Association is that natural radiation contributes about 88% of the annual dose to the population and medical procedures about 12% (56).

Let us pause here and place some quantifying label on what we are talking about. Without some way of expressing dose quantitatively our exposure has little meaning.

The sources and mechanisms of all forms of ionizing radiation is certainly important to the inquiring mind but what we really want to know is the bottom line- how much specific destructive energy is entering our body, over what time period and what risks for illness does it pose.

The radiation literature over the decades has struggled with numerous thoughtful scientific ways with names to provide some rigorous metrics to the field of radiation- the Curie, the Rad, the Rem, the Roentgen, the Gray, the Becquerel, the Sievert - all addressing different views from the bridge.

Through the efforts of the International Commission on Radiological Protection (21) we have finally settled on the Sievert (Sv) to express the safety unit of radiation. It is named after the famous Swedish investigator Rolf Sievert who devoted his life's work to the study of the biological effects of radiation. The Sievert, a derived unit of IR, represents effects such as the probability of cancer induction and genetic aberrations. In short it represents the biologic effects of absorbed IR rather than the physical quantity emitted by a source.

For example a 1Sv fresh exposure carries with it a 5.5 % chance for a human being eventually developing cancer. Doses above 1Sv received over a brief period of time will result in radiation sickness which as the dose increases and can progress to death within a week or two.

The ICRP guidelines assume there is no safe level of exposure for the public so that every reasonable effort should be made to prevent acute exposures and reduce chronic exposure below their recommended regulations.

These guidelines are conservative and contentious among experts. For example as of this date there are only a small number of proven major health hazards from low dose chronic radiation exposure and settings are quite broad for different categories of workers and the cancer statistics bear little relation to low dose cumulative values (57).

They have set an annual all source exposure of 6.2 mSv (thousandth of a Sievert) for the general public, 1 mSv for those under 18 years of age and women during pregnancy; and as much as 50 mSv for those who work around radioactive material.

Airline pilots and flight crew are limited to 20 mSv per year.

(Note - people who live at high altitudes like pilots are less protected from space radiation because the atmosphere is less dense, particularly near the pole as opposed to the equator, and they are exposed 24 hours a day. In spite of their location they have a decreased incidence of cancer and we shall address this later).

By far what can be adding to our recommended yearly dose relates mostly to medical radiologic procedures. These exposures range everywhere from an additional 0.42 mSv for a yearly mammogram (with its justifiable health payoff) to a possible needed whole body Cat-Scan at 10 mSv.

So once again we ask if these exposures add up year after year to some cumulative risk from birth to old age and all living things live in this implacable silent cloud of ionizing radiation how come we are still here to write and talk about it?

The answer has not come easily. Not until the major breakthroughs in genetics and molecular biology have we begun to understand and appreciate the remarkable resilience of natural selection. The fact that a simple cell by virtue of adaptations in its DNA-RNA-Protein sequences and arrangement can under limited circumstances resist, repair, heal and bring injured tissue back to normal function has allowed us to survive over time.

But one question begets another since this process is limited and reaches a threshold between and within living species.

We know that the same dose of IR impacting acutely as opposed to a slow accumulation has a much different outcome. The results of both of these scenarios can differ from one individual to another in terms of time and severity; ie. the impact of an acute exposure may be expressed at once or years later.

The characteristic bad outcomes vary widely in nature from person to person and clearly relate to the cell cycle where disaster may come as the cell divides.

Thus we see deadly effects on tissues that rapidly reproduce themselves such as bone marrow and the gastrointestinal lining.

Does a family history of cancer make one more radiation sensitive and less reparative?

Does the probability of cancer occurrence and or sexual (gonadal) and somatic (body) mutation change over time once one reaches maturity? If so why and in whom? Are there markers for who will respond in a particular way? Does sub liminal background radiation accumulation have any effect at all on later challenges? Why does high dose total body radiation cause different organ outcomes in different individuals? Can one develop resistance to IR similar to acquired immunity? How effective are the anti-IR drugs that might help the repair of injured cells? These are some of the important questions that will require answers.

For the purposes of this book we have to ask a key question.

Does knowledge of this background exposure or intrinsic reparative ability in anyway gives us a leg up as to how man in space will accommodate the much more ominous environment? Is the Earthbound accumulated radiation exposure that an astronaut acquires a factor in his or her resilience or sensitivity to IR in space. And how long should we wait to know the answer?

The message then is that we on this planet are bathed in unremitting sunshine and we have a protective atmosphere that for the most part screens out the solar more destructive output. We live in an aura of background ionizing radiation but nature has provided us with repair mechanisms.

In addition, we have a remarkable magnetosphere (to be discussed) that adds greatly to our protection and we need to understand this gift of nature as well.

Chapter Three
THE DEATH RAYS

We have talked about the broad electromagnetic spectrum that emanates from our star but this is but one phase of solar output. There is a major element that reaches to the far outskirts of the solar system that is a significant player - *The Solar Wind.*

This is a constant but irregular flow of high-speed charged particles that blow off from coronal holes in the outermost layer of the Sun. It is a composite of *electrons* and *protons* seeded with an 8% component of helium nuclei (*alpha particles*), traces of heavy ions and atomic nuclei pouring out from the enigmatic super-hot outer solar area - the *Carona*. It blows off in all directions at 400 million miles per hour and travels to the outskirts of the solar system. As noted, we can only get to see this coronal feeding zone with our naked eye during a solar eclipse or with special instruments.

Make a mental note. This is the principal moment to moment solar system radiation space environment. These particles are streaming off the sun in all directions at the cost of millions of tons of solar mass. It has been doing this since the solar system formed 4.6 billion years ago and is barely noticed by the massive star.

The solar wind has stark variations. It is far from uniform, with changes in speed, density and temperature. It harbors vast magnetic clouds whipping through slower magnetic groups all racing past Earth out into the solar system.

There are many questions to be solved by NASA missions aimed at coronal behavior but the solar wind phenomenon seems to be explained by the fact that the coronal plasma (an ionic soup) is so hot the Suns' gravity is unable to hold onto it. For the most part we are protected from it, however it serves as the conduit for the dangerous radiation to be described.

Which brings us to what seems to be one of the least appreciated of all our environmental gifts, a relatively late bloomer in astronomical planetary understanding - The Earths' *magnetosphere*.

This is a magnetic shield surrounding several of our solar planets. We have bragging rights for ours is the most efficient and powerful of them all. Its genesis comes from a swirling flow of molten iron deep in our core that creates an umbrella of a magnetic field surrounding Earth whose frontal shield reaches out some 36,000 miles into space.

Housed within this protective barrier at the level of our equator is a massive doughnut shaped invisible structure with an outer layer that principally traps the solar wind electrons, an inner layer sequestering solar and dangerous cosmic protons and an in between layer yet to be fully defined.

This complex structure first detected in 1968 by Professor James VanAllen is called the *Van Allen Belt* and is activated and expanded rapidly to thousands of miles into space in response to impacting solar storms.

Recently discovered is an impenetrable wall lining the inner margin of the outer belt that resists highly accelerated "killer electrons" heading our way. All space activities that transcend low earth orbit are planned to minimize passage through this important deadly but protective barrier.

The Apollo Mission whizzed through this dangerous matter without event and the hope and intent is that all deep space adventures will do the same.

Studies show that the solar wind smashes into us at supersonic speeds passing through a bow shock compressing the magnetosphere, feeding charges into it and for the most part slipping off in all directions into the belts or space. We are spared the full impact of their particulate energetic rays and life continues on Earth.

We have then set the scene of a beautiful planet bathed in sunshine, wrapped in a protective layered blanket of atmosphere and an almost magical magnetic wall all seemingly designed to keep life continuing on its remarkable path, a thriving plant and animal life that can mend itself even when exposed to its own generic radiation.

This would have all the characteristics of an astronomical fairytale if the story ended here but as fate would have it there is a very different tale to be told.

Simply put, our friend the Sun is not always friendly. *Sunspots* have a long and familiar history. They have been observed, mapped out and recorded for centuries. They are actually polarized magnetic twisted hot spots on the photosphere (the visible surface) but are relatively cooler than the local environment so they appear dark with outer gray (penumbra) and characteristic central blackness (umbra). They are harbingers of the Suns' activity and though for centuries have always been called "spots" they are Earth - sized at a minimum.

The Sun is on a continuous loop of activity over time, *an eleven year cycle* in which sunspots are the markers that can disappear for days at the beginning of the cycle in contrast to the end of the cycle wherein sunspots are larger, more numerous and appear in broad paths. All of this appears to influence in some way our weather patterns, with the active part of the Sun cycle associated with greater disturbances.

What seems clear is that the Suns' incredible magnetic activity due to massive moving charges create linear fields that get twisted broken and rejoined in complex patterns that bring about these changes.

However when astronomers talk about the relatively "active Sun and the quiet Sun" they are talking about a lot more than sunspots They are talking about some concomitant major league changes in the Suns' behavior. They are talking about phenomena that are the origin of three of the five sources of potentially lethal radiation.

There are essentially two primary events that are the spawning grounds for these nasty emanations. Number one is a phenomenon known as a *Solar Flare*. Number two is known as a *Coronal Mass Ejection* (CME).

Solar Flares are sudden powerful brilliant brightenings discharging powerful releases of high frequency electromagnetic energy and charged plasma into the solar wind, occurring episodically anywhere over the Suns' surface with energy outputs in the order of 160 trillion million tons of TNT, lasting minutes to hours, the advanced EM party travelling at the speed of light. Some are more powerful than others but all are major happenings. These flares again are associated with the tremendous heat that develops as twisted ruptured magnetic lines reunite and generate their destructive rays in the immense heat of local plasma. They appear to connect to the photosphere, where the sunspots hover, up through the chromosphere and out through the corona with gigantic pillars of ionic particles which are released by the flares energy.

Solar Flares though often lasting only minutes can occur several times a day during sun spot maximum and at the 11 year low often quiet down to one or two per week. What is important to us is not only their frequency but their direction. Heading for us even with all our built in protections they constitute the solar storms that can overwhelm the magnetosphere, ionizing the upper atmosphere, disrupting short wave radio, air flight communications, dragging down orbiting satellites, damaging our electrical grids and ending up with spectacular auroras at both poles.

Although the bleak biologic implications of impacting with Solar Flare particle- ion output in space has been recognized since we first put astronauts out beyond our atmosphere not until recently have we been giving full regard to the other important feature of this Phenomenon - namely the intensity of the associated electromagnetic spectrum and the lethality of the often accompanying Hard X-Rays from an X rated (highest class) Solar Flare that creates a somewhat unique type of tissue damage but nevertheless can lead to cell death (44).

And what is so dangerous about Hard X-rays is that you would get only minutes to scramble to shelter at the first warning of a major Solar Flare if you are walking around in space in your space suit, whereas the bulk of the associated fast moving particle charges take one half hour to over 24 hrs. in order to arrive in the locale. Hard X-rays travel at the speed of light and unless you are close to some kind of shelter you are going to get zapped through your space suit.

Hard X-Rays unlike ordinary X-Rays (<10 KEVs) can carry energies greater than 100 kilo-electron volts and at this intensity can produce serious biologic injury.

But the darkest story of our bright Sun is not finished for their remains the apogee of the solar ray noxious menu – The Coronal Mass Emissions - what astrophysicists call CMEs.

Although the taxonomy of CMEs and Solar Flares can be somewhat confusing, for example when large loops of plasma of varying temperature called prominences rise to different heights from the Sun's surface and leak off charged particles (flares) or break off releasing clouds of charged particles they may be difficult to classify.

At one time Solar Flares and CMEs were thought to be the same phenomena but today we know they are a distinctly different breed.

For the sake of clarity we are mostly centered on clouds of the mysterious enormously hot coronal plasma that can suddenly blast out of the coronal locus carrying with it a billion tons of charged mass in one spectacular explosion, traveling in and shocking the solar wind at speed that take several days to arrive at targets out in space. When aimed at our planet these CMEs cause geomagnetic storms whose extremely hot source result in charged particles at the top of the scale. They penetrate our magnetosphere and like Solar Flares disrupt radio and electrical transmissions, can damage satellites and other electronic equipment and cause severe economic losses due to sustained power outages wiping out our grids.

CMEs are often associated with Solar Flares however as noted they are clearly distinct, and when generated from the super-hot corona as opposed to the photo and chromosphere are more powerful and therefore more destructive.

CME frequency is similar to Solar Flares, roughly 3.5 per day during solar max and about one per week during minimum and typically travel the 93 million mile trip to planet earth in up to 5 days. The challenge to these death rays is to be able to predict their outbreaks and direction in space and get adequate shelter if needed.

Recent research at the Goddard Space Flight Center by the Gopalswamy team (16) have recognized CMEs crashing into slower charged plasma in the solar wind generating the especially dangerous "radiation storm" giving off telltale radio warnings that may prove invaluable as alarms for astronauts working outside of shelters in space.

Embedded in the Solar Flares and CMEs are unique Solar Particle Events (SPEs), intense agglomerations of positive ions, the "proton storms". These are highly energetic marked accelerated particles caught up in the solar surface by exploding Solar Flares or in shock waves of CMEs impacting in interplanetary space and include a minor assortment of helium ions or higher mass ions. They are part of the spontaneous episodic solar hazard to spacecraft and astronauts.

In October and November 2003, two years after solar maximum the so called "Holloween Storms", several billion tons of electrified gas travelling at high speeds past Earth and Mars, the highest intensity of solar events ever recorded, blasted off unexpectedly from the

Sun and within a few days knocked out detectors as far out as Saturn.

Capricious and unpredictable, these solar events are an ominous threat to adventures in outer space.

Which finally brings us to the grand-daddies of all lethal ionizing radiation the Cosmic Rays. To understand them we will be turning our attention now to regions far beyond our solar system.

Chapter Four

THE VILLAIN FROM OUTER SPACE

So far we have been writing about ionizing radiation distributed throughout our solar system whose heat sources and consequent energies are enormous but limited; an environment of charged particles (protons) borne by the solar wind, trapped by Earth's magnetic field in the Van Allen Belts, and carrying intermittent indeterminate explosions of high energy deadly hard X-Rays and particles of Solar Flares and Coronal Mass Emissions, the so-called Solar Particle Events (SPEs).

We now must address a type of often labelled mysterious ionizing radiation (IR) of sources that dwarf our solar system, rays from outer space that continue to be under careful scientific scrutiny for there is much still to be learned about them- the Galactic Cosmic Rays (GCRs.)

Here is brief look at what we know about them.

GCRs, unlike electromagnetic radiation which travels in straight paths, consist principally of very high energy charged particles that have such a hectic rebounding travel path as they wing through deep space bouncing off astronomical debris in all directions. The best astrophysical analysis has difficulty in determining their exact origins though recent researcher have identified supernova remnants as one of the important accelerators.

Working backward from the secondary rays that shower from cosmic ray impact they appear to be also coming from the immense heat of a mix of deep space early supernovas, "active galactic nuclei", spinning neutron stars, the flash jets around back holes and possibly gamma ray bursts.

They are a vast collections of high energy fully charged protons (of electron stripped elements) with a ratio of hydrogen 85%, helium14%, and 1% for more complex ions up to iron, and heavier periodic table elements, the HZEs, (H for high charge, E for energy, Z for the proton number) with a spread of energies of about 12 orders of magnitude within the collection.

Professor Angela Olinto (35) of the U. of Chicago, a world expert on HZEs tells us the highest energy particles pack a trillion times more punch than the lowest energy ones. They transgress the walls of space ships and endanger all personnel so there is no place to hide.

These energetic charged particles are not only powerful enough to destroy human life but also can destroy the microelectronics of instruments in space so that telescopes and detectors of every ilk are vulnerable.

GRAY: *Radiation energy is expressed in electron-volts (eVs)- the energy given to an electron by accelerating it through 1 volt of electrical potential difference.*

HZEs max out around 3×10^{20} eV. (That's 40 million times the output of The Large Hadron Collider!). Even the average Cosmic Ray is a whopping 4×10^{11} eV.

Compare this with a medical X-Ray of 0.2 MeV and Gamma Rays around 3Mev.

Paradoxically this implacable intermittent storm of GCRs increases during solar minimum with an intensity that decreases somewhat during solar maximum. The solar wind activity bearing definite magnetic shielding of its own tends to modulate the intense flux of GCRs. Space flight planners recognizes this and are likely to make changes to fit this important relationship.

GRAY: Even during solar minimum there are unpredictable Coronal Mass Ejections and Flares that can result in a decrease of 30 % of incoming Galactic Cosmic Radiation the so called Forbush Decrease. This occurs 24-48 hours after a Solar Particle Event and at the same time that we experience the magnetic storm and consequent auroras. The precise cause is under much study.

On Earth we are bombarded by a constant shower of secondary cosmic radiation attenuated by our atmosphere that rains down upon us 24 hours a day and for the most part passes right through us harmlessly.

However, when outer space Cosmic Radiation smashes into solid material such as the wall of a space ship it generates a conglomeration of secondary rays that can damage tissue.

Space secondary cosmic rays can be dangerous and destructive. The output is mostly *muons*, (which are heavier than electrons and carry greater energies), electrons and neutrons (which can have high kinetic energy, penetrate and generate ions). Secondary cosmic radiation can also generate an assortment of x-rays, protons and alpha particles (electron stripped helium), antiprotons and subatomic particles that vary with altitude and location.

The HZEs are a small but important class of GCRs that can drive through everything except deep lead. They occur less frequently (about one or two a month) but are clearly the most threatening of all space ionizing radiation. All Cosmic Rays travel at high relativistic speeds with the high mass rays travelling close to the speed of light.

Yes, atmosphere attenuated secondary GCRs pass through our bodies every second of our life with little or no biologic impact. But, out in space this is a different story indeed where astronauts have to contend with lethal primaries and the secondary showers of subatomic particles that rage into them from cabin and vehicle shielding and through their space suits.

It is not surprising that all Cosmic Rays are potential biological anathemas but the HZEs, though in much reduced numbers, with encounters only after many weeks of exposure in space, are the most destructive of all for every kind of life.

The metaphor would be-ordinary Cosmic Rays hit like bullets, HZEs hit like freight trains. Ordinary high energy protons are hydrogen ions, heavy HZEs can be mass nuclei up to iron-with a consequent +26 charge and as noted rarer higher periodic table values born in the extremes of supernova explosions can appear.

It is worth repeating. Their occurrence is such that they may show up on detectors only one or two per month. But stay in space over time as planned and they will find you. It is the hallmark statement of this book.

Although scientist have at last invented accelerators that can produce unprecedented energies to unlock nature's secrets, such as the Large Hadron Collider on the Swiss-French border, these values are far inferior to what we have coming from disturbances in our galaxy. Whatever phenomenon these far off gigantic disturbances induce they are unmatched accelerators.

We have come a long way since Victor Hess noted that his electroscope went into a frenzy as the balloon in which he was traveling ascended into the upper atmosphere in 1921. But we have yet to learn all the lessons these powerful rays have to teach us.

Astronomy is a science of evolving surprises so that today's truth is tomorrow's correctable. Even as we hover over the awesome power of cosmic radiation there are new whispers from outer space of a new high energy source.

In November 2013 The University of Wisconsin's Ice Cube Telescope at the South-pole in a vast one cubic kilometer of ice discovered two of the highest energy neutrinos ever observed, (1.14 petaelectronvolts -$10^{15\text{th}}$) impishly named Bert and Ernie (of Sesame Street) and have followed up with recording 28 more candidate high energy neutrinos (55).

These are not the showers of low energy mini mass neutrinos that rain through us every day. These are extra-energetic galactic neutrinos probably spun off by cataclysmic events in exotic explosive clashes of the early universe and they open up a new world of extra-galactic astronomical investigation.

As of now these appear to be rare events observed over many months but from our standpoint raise the inevitable question: Could these extremely high energy almost massless zero charged rays have any significant unrecognized biological impact?

As of now Cosmic Rays are our biggest radiation threat in space. They have the most imposing presence, are the most unrelenting and resistant to shielding and no space initiative can ignore them.

Chapter Five
PLANS TO RETURN TO THE MOON

Are NASA astronauts going back to the Moon or not? There is understandably some public confusion over this matter.

In the Authorization Act of 2005 NASA was instructed "to develop a sustained human presence on the Moon." Thus began a new ambitious Constellation Program to send astronauts first to ISS, then to the Moon, and then to Mars and beyond.

NASA made it clear on its website that the Moon project on its own was an important component listing six reasons:

1. To extend human colonization
2. To further pursue scientific activities intrinsic to the Moon

3. To test new technologies, systems, flight operations and techniques to serve future space exploration missions
4. To provide a challenging, shared and peaceful activity to unite nations in pursuit of common objectives
5. To expand the economic sphere while conducting research activities that benefit our home planet
6. To engage the public and students to help develop the high-technology workforce that will be required to address the challenges of tomorrow

This was glorious rhetoric but when in 2011 the Obama administration had to implement major budgetary constraints the Constellation program was gutted and the Moon program was cancelled.

However this was not to be the end of the story. There has been in the background a constant drumbeat by political, scientific, and industrial forces to reconsider the options.

But NASA's unrelenting focus on the Mars initiative according to administrator Charlie Bolden as of 2013 was that a man on the Moon lander costs were prohibitive and NASA would not be putting a man on the Moon.

However, the following year we hear that human presence on the Moon could be a requisite for the Mars or Asteroid project and must be considered.

Putting humans back on the Moon simply will not go away. The Russian Federal Space Agency announced they will definitely do it by 2030 (43).

GRAY:

Astronomically stated the Moon is Earth's one and only natural satellite; most planets have more than one. Relative to our size it is the largest Moon of the 173 circling planets in our solar system. We think it formed one or two billion years after Earth formed when a Mars sized object smashed into the still nascent Earth and knocked a large mass of planetary debris out into neighboring space where it took shape.

The Moon's distance from Earth averages around 238,855 miles. Its diameter is 2,160 miles (27% of Earth's). It has a little over 27 days of axial rotation as it orbits Earth so that it always shows us the same face. Our gravity keeps it in orbit and its gravity creates our ocean tides.

Trip time (exposure to radiation time) to the Moon will always be an important consideration. Apollo missions took an average of a little over 3 full days to negotiate the 380,000 km. distance. New rocket and launch capabilities are sure to shorten this time. NASA's New Horizons Pluto Mission sped past the Moon in 8 hours and 35 minutes as it headed out.

If you prefer to take the slow route such as the ESA SMART-1 lunar probe using the dramatic fuel saving ion engine (using only 82 kg. of Xenon!), it will take a little over 1 year to get there, which would include a serious interval for radiation exposure.

....

From 1963 to 1972 the dedication, bravery and engineering genius imbedded in the six Apollo landings on the Moon brought about a series of space events that are now enshrined in American History.

Twenty four Apollo astronauts are the only humans to date who have left earth's sub-orbit beyond the magnetosphere.

Twelve astronauts with six space flights spent a maximum of 160 hours on the Moon outside of the Lunar Module. Everyone in the entire program spent less than a week in outer space. We in turn planted our flag all over the lunar dusty surface.

Space weather experts point out that no major solar event took place while our Apollo teams were deployed.

Space weather experts also point out that the Apollo mission narrowly escaped the "big one" that did occur in August 1972, one that could have been devastating. This legendary storm occurred between Apollo's 16 April fight and Apollo's December flight.

Dr. E. Cucinotta, one of our most prominent experts on space radiation, estimated the 400 rem (~4Sv.) absorption from that storm would have overwhelmed any "Moon walker" and precipitated the life threatening "acute radiation syndrome" (possibly mitigated only by a quick trip home and urgent medical rescue). This abrupt unpredictable incident would have changed the entire Apollo celebration (30).

The health follow-ups on this important group will be discussed in chapter 9. In general they are quite encouraging. Our statisticians would find any conclusions based on this small sample and trial time "utter nonsense". But, let us face the fact that there is no way to frame a randomized double-blind placebo controlled test plan around the confounding question of the health hazards of deep space.

One interpretation of this laudable data for Moon enthusiasts can mean we now have a free pass to revisit the Moon with men and women, plan exploration in detail, exploit whatever hidden treasure we can find or conger up and create a thriving settlement.

NASA's Constellation program faded under the pullback in government funding but the hopes and plans have remained with many enthusiasts lobbying with such catchwords as "Habitats", "Power Stations", "Greenhouses", "Prospecting", "Mining" and "200 glorious days".

J. McDowell at the Harvard-Smithsonian Center for Astrophysics notes in an interview with The Telegraph that the recent lunar orbiting missions have produced huge amounts of detail that have breathed life into the hope of putting a man back on the Moon.

One scenario reads to first establish "outposts" to serve as stations for the Mars mission or an Asteroid visit with larger crews rotating for up to six months. The locus may well center on the moons' South Pole where there appears to be ample sources of hydrogen and water (ice). All of this is capped by great improvements in rocketry both in terms of safety, convenience, and speed.

Ken Murphy on the internet adds to NASA's six reason listing "25 good reasons" why we should go to the Moon, all part of the cheerleading from the bleachers.

In a recent article by Buzz Aldrin (the second man to plant his feet on the lunar surface) he notes that recent international robotics have verified that the Moon is a mother-load of useful materials.

Moon proponents point out that compared to Mars and Asteroid flights the Moon is literally next door with possible emergency services ready to go back and forth in less than a week so that with telemedicine for diagnostics an astronaut with an acute appendix might get home before rupture and peritonitis or get their osteoporotic fracture repaired in time.

It all sounds so good that there is little concern now about the open season for private enterprise to incrementally enter the deep space arena. NASA has encouraged some of this with its ISS working partnership.

There was commercial excitement upon effective servicing of the International Space Station (remember skeptics said it could not be done). Now the great American corporate know-how is getting ready for whatever economic advantages there will be 240,000 miles from the company's front door and those with the nerve and resources intend to bring it about.

Quietly in the wings The Golden Spike Company led by Gerry Griffin and Alan Stern announced in December 2012 they are aiming to provide a commercial opportunity for people to set foot on the lunar surface as early as 2020. It intends to raise $7-8 billion dollars with Northrop Grumman developing the lunar lander.

Dodging the Death Rays

Unlike the Apollo Mission the orbiter and the lander will be on separate rockets. The plan is to first launch a waiting landing vehicle into Moon orbit followed by a crew launch with two paying customers. There is a detailed plan for Moon landing exploration and a return splashdown landing all at an estimated cost of 1.5 billion dollars per trip. As of 2015 Golden Spike has not backed off (16).

It all sounds like an exciting center stage adventure with any negative factors a rude intervention.

We note little mention of the abrasive lunar dust that Apollo astronauts complained about, or the severe temperature changes -387 Fahrenheit at night to +253 during the day, the vast scenic desolation, and the unrelenting problems on the human skeletal, cardiovascular and muscular system in a world of reduced gravity; the extended acquisition problems of oxygen, food and water.

But our concerns remain to focus on the environmental problems of ionizing radiation, that radiation from predictable high energy Cosmic Rays and unpredictable dangerous Solar Particle Events and the radiobiological events that could ensue.

The Moon has no protective atmosphere and no protective magnetosphere, and everyone engaged in spacewalking there is a sitting duck out in the open, a challenge to the new radio-protective chemical suits under development. There is the ever present demand on anyone involved in extra vehicular activity to pay close attention to the solar alarms to make sure that one of the daily or weekly SPEs are not heading their way. And of course if one is to take off that suit they had better be in an appropriate shelter and hope that HZEs are not finding them.

Therefore putting human beings back on the Moon for an extended period of time will never be a slam dunk and there are sober voices that must be heard.

These sudden major unpredictable storms can occur independent of the solar cycle, such as the Hard X-RAY type traveling at the speed of light, and highly charged protons with mass moving close behind. Simply put, what can bring down a continental electrical grid can potentially destroy a human being in space.

In response to one of the bigger sudden proton storms on record in January 20, 2005, an X Class (the top) Solar Flare hurling a billion-ton cloud of charged particles into space NASA scientists note (as in the August storm of '72) that anyone caught outside would have little time to get to shelter. This storm was calculated to produce a >0.5 Sv storm in absorbed IR. An astronaut roaming about on the Moon could suffer the nausea and vomiting of acute radiation sickness and undergo the inevitable damage to cells and tissues but note that this was the final storm emanating from the locus of a sunspot labelled NOAA-780 that had generated 4 previous Solar Flares the previous week! (30).

The point of course is that high dose Sieverts add up to lasting damage and note that the ever present Galactic Cosmic Rays have not even been factored in.

The most optimistic planners for Moon return agree that any "early colonists" facing the harsh environment without much spare equipment will have to go underground. The scenario is certainly not user friendly. These pioneers would have to labor always at great risks out in the open and resolve the hazards by living in caves.

Finally there has been increasing recognition in this last decade of the obvious hazards of GCR and SPE radiation impact on lunar soil (47).

One of the surprising findings that emerged in 2009 from NASA's Lunar Reconnaissance Orbiter (LRO) was that Cosmic Ray secondary radiation emanating from lunar soil was a potential bio-hazard that needed to be examined. (LRO)

Recent Moon research by The University of New Hampshire scientists using the DoSen Detector has detected a previously unrecognized and possibly significant additional source of Neutron Radiation created by high energy Cosmic Rays impacting the Moon with its lack of atmosphere or magnetosphere and smashing into the lunar soil. Neutron Radiation by definition as we have noted is not charged but as soon as it penetrates any mass it generates energetic ions that are highly destructive to human tissue.

This "splashback" of Neutron Radiation was not noted in the Apollo explorations and will need study and confirmation. It has been referred to disarmingly as the "radiation albedo". It could represent a new and perplexing problem for shielding and one more hazard to defend against over time.

What exposure limits this will add to the already multifactorial radiation facing our astronauts is yet to be determined but it is clear that those who go to the Moon are going to face IR from all directions (14).

In summary, it should be noted that unlike the International Space Ship (ISS) which is well within the protective magnetosphere the Moon enterprise is out in unprotected space so that human endeavors on a return to the Moon will be constantly challenged from episodic solar particle events and the ever present high energy display of Galactic Cosmic Rays with exposures over time that must be recognized as a major health threat. There will be many strategies developed in an attempt to neutralize these warnings, hopefully before we have a major human presence on this desolate world.

Sometimes delay and not time is of the essence.

Chapter Six

THE LONG MARCH TO MARS

The public's long love affair with our nearest planet Mars, continues with un-abating zeal even though the presence of man-made canals (Percival Lowell, 1906) and enemy aliens (Orson Wells, 1938) were bogus and the reputed fossil recently recovered has failed to demonstrate a once living organism.

Staunchly behind the public's adventurous interest has been a steady upbeat professional scientific interest that was fostered by the realities of the Apollo program and has generated, over the years, a continuum of Mars Mission plans from a variety of national and international organizations and space agencies.

Mars, the terrestrial red planet, is our nearest neighbor in astronomical terms. Only some 35 million miles away, it has striking similarities that tend to obfuscate the differences.

Mars, the remarkable red planet, (red because of rusty iron oxide that dominates its surface) rotates as we do in only a fraction more than our 24 hour day. Its axis of inclination unlike the inner planets is similar to ours and because of this inclination it has seasons in relation to the Sun not unlike ours.

Mars has volcanoes, mountains and valleys such as ours. It not only has a moon, it has two. Some pictures of its surface remind us of the New Mexico desert. It has been labelled our twin, our sibling, our astronomical cousin. It resemble us in so many ways the conviction of many Mars enthusiasts is that it must have somehow contained life similar to ours and we should make every effort to confirm this. There also is considerable interest in taking residence on its surface with both temporary and permanent installations in mind.

What Mars does not have is an accommodating temperature for human activities (average daytime -71 to -53 degrees F., nighttime -184 to -85 degrees F.) with its further distance decreasing its sunshine effect by over 40 per cent. Whatever water may exits is deep under its surface or locked up in polar permafrost. It has 38% of our surface gravity and when its surface suddenly disappears in hurricane-like massive clouds of sand and lava dust its surface changes in a vista of craters and dunes.

Dodging the Death Rays

These dust storms can be violent and unpredictable with wind speeds exceeding 125 miles per hour, the dust sometimes not settling for many weeks. This in turn can create a loss of solar power with the domino effect that can threaten oxygen extraction, water acquisition, greenhouse requirements etc.

NASA has given ample attention to this fine invasive dust in order to protect all of its instruments with good results. What we don't know is how an astronaut would hold up in the reduced gravity of a Martian gale.

Important to our subject is the Martian rarefied atmosphere which the charged solar wind continues to eat away. This thin atmosphere is 96% carbon dioxide, with argon and nitrogen filling the remainder. It has only traces of oxygen that cannot now sustain life. Astronauts as noted will be required to manufacture oxygen from water, and the water will need to be extracted from the soil or the polar ice.

Mars atmosphere will therefore not make a significant attenuation of incoming higher energy ionizing radiation (IR). Just as significant, the planet lacks an effective magnetosphere which for starters means an open season for IR in any unsheltered area. So once again we have an environment where shelter is king.

GRAY: *An interesting question arises from the possible contamination of astronaut shelters with fine blown dust of the Martian surface upon reentry from outside extravehicular activities. If the air-borne soil has undergone significant spallation that takes place in the soil from unshielded exposure to galactic cosmic radiation, as would be expected in this environment, it would literally envelop an astronaut's spacesuit with traces of a variety of light elements and an assortment of cosmogenic nuclides such as tritium, and isotopes of aluminum, carbon-14, chlorine, iodine and neon. Bringing this background radiation contamination into one's Martian abode could build up over time. It may well be just one more radiation problem for NASA problem solvers to solve.*

With all of this in mind the public, (who will pay for it all, depending on which scenario with cost estimates from 6 to 500 billion dollars) has a right to asks why even go to Mars?

The answerers make the following points.

1. The inevitable pull of the unknown, the prospect of adventure and discovery. It is the drive that has brought about all of the precious geographic and resource values of our world.

2. The thrill of knowing that science and engineering has not lost its amazing genius.
3. It promises to jumpstart developments in technology, in areas such as recycling, solar energy, food production and medical science.

That these reasons are shared world-wide is confirmed by the impressive number of proposals for "Human Missions to Mars" that began with rocket scientist Werner Van Braun in the late post-war 1940's and have continued with numerous additions, modifications, inventions, and interventions to date. Russia, France, Netherlands, China and India all have proposals on the table some of which have launch dates preceding ours.

There is a litany of different types of launch vehicles proposed, technologies for survival, different numbers of months of travel time, different crew numbers, and a broad extension of days, months or years to be spent on the planet.

Some ventures are simply interested in the astro-engineering feat of landing and getting out safely, others in extensive exploration, still others in a total commitment for settlement and an intention to "transform" the planet.

There are plans to bring in shifts, every conceivable type of supply, building material and plant life. There are "one way" plans of going and not coming back. The variations and permutations are a testimony of continuing serious aspirations and intent.

The International Space Exploration Forum with representatives from 30 countries met in Washington on January 2014 to underscore the importance of space exploration, its value to humankind and the need for international commitment and cooperation (U.S.), so it is not impossible that the ultimate mission will be a joint international effort sharing the risks and the costs.

And now once again private enterprise is planning to enter the race, and a race it appears to be, since they are talking about going to Mars in the 2020s.

"They" are the people such as at Mars One, a Dutch non-profit foundation, the mother company of the for-profit Interplanetary Media Group securing funds from its investors.

In a highly publicized search for people interested in going to Mars and not coming back they have selected 100 credible candidates out of an amazing round of 202,586 applicants. A selected 100 were culled by interview from a filtered group of 600 by Chief Medical Officer Norbert Kraft resulting in 50 men and 50 women principally American and European, all of whom understand the risks and needs for team playing and motivation.

This one way mission is headed up by Dutch entrepreneur Bas Lansdorp. The project has the clear intent of sending waves of 30 aspiring men and women to live and die on the planet. These cadres will arrive every two years with contraceptive supplies to avoid any pregnancies in their early colonial period!

The drumbeat of national and private enterprise in space is on the march with little mention of those inconvenient death rays.

In May of 2013 NASA (in its ongoing concern about radiation effects) sponsored an important study by Zeittlin, Hassler, Cucinotta et al.,(59) published in the prestigious journal Science in May 2013.They measured energetic particle radiation in a 253 day transit to Mars on the Mars Science Laboratory spaceship. The instrument was turned on 10 days after launch and turned off three weeks before landing.

A radiation detector made detailed recordings inside the spacecraft and was placed on the shielded rover Curiosity. It found the dose equivalent for even the shortest round trip and comparable shielding was 0.66 +/- 0.12 sieverts. This exceeded the estimated life time cancer deaths limit of 3 per cent that NASA has set for males and female astronauts. It is the most definitive study of its kind to date.

The study only reported on sheltered data while on the spacecraft and did not include the exposure on a stay on Mars surface. Solar activity was at a minimum but five solar events increased the daily exposure rate by 13.4.

A subsequent article by the same team recording the detector data on a Curiosity's sojourn on Mars surface brought the total mission dose of ~1.01 Sv for a round trip Mars Surface Mission with 180 days each way cruise and 500 days on the Martian surface all during an active solar cycle (which you may remember allows for less Galactic Cosmic Ray exposure).

This value, if correct, speaks for the importance of recognizing the health risk of extended space flight as proposed in all Mars mission scenarios both scientific, exploratory and inspirational; and it speaks for collecting more confirming data.

NASA's concern about space radiation goes back to the inception of our human spaceflight programs. All crew members of Project Mercury ('59-'63) had extensive detector study and follow-up.

There followed the studious radiation concerns of NASA's Jerry Modisette et al. high-lighting the radiation precautions needed for the Apollo Moon Mission using detectors and the Solar Particle Alert Network and the need to proceed quickly through the VanAllen Belts which had been discovered in 1958.

This roll call of caution has continued to date with the now added recognition of the hazard of Cosmic Radiation.

One of the most critical outspoken experts in this field has been NASA's own Dr. Frank Cucinotta formerly of The Space Radiation Health Project at the Johnson Space Center.

He first published his serious concerns in the journal Radiation Research in 2001 (10).

In a telling interview in Science @NASA in 2004 he again focused on the cancer hazards in a trip to Mars with unacceptable values in male astronauts (with females twice as vulnerable). The bottom line being there could be no human trip to Mars without solving the problem of spaceship shielding. As of 2015, with the Zeitlin team, he has not relented (59).

What will the new polyethylene spaceship material do to change the picture? What will new data acquired during a different period of the solar cycle tell us? What new propulsion mechanisms will move humans through space more expeditiously?

We await the answers.

Chapter Seven

THE ULTIMATE NASA MISSION

It is not Sci-fi hype. The fact is there is a real existential threat out there that has the ability to wipe out all life on earth. There are two of them to be precise. They have tried to destroy all life on this planet in the distant past and we have every reason to believe they will try it again sooner or later.

"They" are, of course, the ever present swarm of city sized Near Earth Objects (NEOs) whose orbits can potentially cross ours, those flying masses of solid carbon-iron-rock missiles and the expected and unanticipated massive swooping ice balls knocked out of their distant paths by some errant passing star - the Asteroids and their kissing cousins the Comets.

....

Dodging the Death Rays

GRAY: *A quick review of the nature and place of this non-planetary junk in the solar system may be in order for you science buffs.*

The measuring stick for the solar system is called the Astronomical Unit (AU) and is the distance of planet Earth from the Sun (approx. 93 million miles).

Comets, the dark crusted giant balls of compacted planetary dust and ice reside principally in two areas.

Billions of them reside in a vast spherical reservoir called the Oort Cloud located from 50,000-100,000 AUs (1AU= Earth to Sun distance) out beyond the outer reach of the main solar system. These contain the long period Comets that get knocked out of their locus by the gravitational shove of some(random passing star.

A similar but less dense collection reside in a structure called the Kuiper Belt extending some 10,000 AU out from around Pluto and orbiting around the Sun in an orderly manner along with the rest of the planets. The short period Comets (less than 100 year cycle) that occasionally break out and orbit toward the Sun (and us) come from this belt. We are not sure what provokes this.

Asteroids are solid mostly stone and iron rocks that hover in a belt between mars and Jupiter and probably represent the leftover debris of a failed planet. They are gigantic city sized objects or larger and unlike the far off Kuiper and Oort comets are readily seen, labeled and tracked by both professional and amateur astronomers with their telescopes.

These rocks get smaller and smaller as they get closer and closer to the inner planets, including us, and circulate as boulders down to grains of sand. We refer to them as meteoroids and micro-meteoroids.

When these objects streak through our atmosphere they are "meteors" (the shooting stars) and when they land everywhere (about 100 tons a day) they are the "meteorites".

Meteoroids travel about 22,000 miles per hour and if they smash into a spaceship or an astronaut working in space it is an obvious disaster. NASA does its best to prevent this deadly collision from happening by diverting space travel according to warnings from the Canadian Meteor Orbit Radar (COMOR system) which keeps a constant eye on meteoroid collections in space. It is just one more hazard in space flight that requires attention.

....

Over a half million space rocks (silicates, iron, nickel, carbon) have been noted in the Asteroid Belt by astronomers and over 15,000 given specific names. It would be great if these swarms of space slag stayed safely in their orbits between Mars and Jupiter or out beyond Pluto like their big brother planets. But this is not the case.

In June 2014 NASA announced that it was developing a first ever Asteroid Redirect Mission (ARM) to identify and capture a near Earth Asteroid with an ingenious robotic maneuver and contain it and place it in a stable orbit around the moon. This will entail sending out a spacecraft capable of nuzzling up to a speeding Asteroid, capturing it and bring it to a "distant retrograde Moon orbit" where astronauts aboard an Orion Spacecraft will emerge to explore it in the 2020s. Six promising candidate masses have already been identified with observations of their velocities, orbits and spin.

Importantly, this initiative includes an Asteroid Grand Challenge "designed to accelerate NASA's effort to locate potentially hazardous asteroids" and demonstrate robotic defense techniques to deflect the dangerous ones and protect earth (31).

NASA'S Near Earth Object Observation Program has already catalogued over 1000 of these special NEOs.

Estimates are that random Asteroid invasion is five times more likely than Comet invasion and statistically both are rare happenings. But simply put "random" means what it says and when a Near Earth Object orbit crosses close to ours we have the all too possible random scenario for massive injury or extinction.

The Yucatan crater that appears to have extinguished much animal life and the famous well-preserved Meteor Crater in Arizona are ominous examples from the distant past.

Doubting Thomas folk need only a spectator's visit to Tunguska, Russia where in 1908 an out of earth object flattened 770 square miles of forest releasing 3-5 megatons of energy.

Better yet have them visit the Russian city of Chelyabinsk in the Ural Mountains where shortly after 9 A.M. February 15, 2013 a meteor fireball travelling 11.6 miles per second created a terrorizing shockwave exploding at an altitude of 76,000 feet, collapsing roofs below, shattering over 100,000 windows, with 1500 people requiring medical attention. These of course are only trivial teasers as to what a town sized incoming mass could do to our planet.

We have made the case for the fact that space flight and settlement have irrevocable intrinsic hazards from ionizing radiation that cannot be ignored. Putting human beings under its particular risks is a reality that must be considered for any future plans.

But now we have a decision to make. Given limited resources if we have to take real but unquantifiable human risks should not Near Earth Object Exploration take precedent over all other human space plans?

A supporting argument runs that pioneers have always taken risks and a good argument could be made that if our space pioneers have to take risks let's make sure it's for this unequivocal important goal and not some vacuous space enterprise.

But there will continue to be advocates out in the public forum who continue to endorse an Asteroid Program only if it serves as a jump start for the Mars mission. It will be difficult to change their mind.

Will the radiation risks be any different already covered in Moon and Mars projected plans? As noted the answer is yes they will probably be somewhat increased. Travel time as designed may be similar but sheltering from solar activity and Cosmic Radiation could be more difficult.

Astronaut - Asteroid explorations will require innovative protection. There will probably be no caves to hide in but we may find evidence to the contrary.

In short, the Asteroid retrieval schemes are now in place with plans for a launch as early as 2017 with an astronaut participation by 2021. The President and Congress fortunately are well aware of the importance of this initiative and some of the funding is already provided.

As for Comet exploration, it is already well on its remarkable way.

In early November 2014 the European Space Agency in partners with NASA announced that the Comet Mission's Rosetta Probe had caught up to Comet 67P, 310 million miles from Earth. Within days it was orbiting it and successfully deploying its lander, Philae, which bounced on the comet, harpooned its surface on November 12th, taking detailed pictures and submitting a roster of unique data, this all being a space history making robotic triumph for astrophysicists and astro-engineers.

Again the blueprint calls for eventual capture and deployment where it can be further studied. When, where and how astronaut participation will be exactly woven into such missions remains on the planning board, but participate they will. Space Expert Sophie Allen at the National Space Center tells us this will be difficult but feasible.

The bottom line to all of this is that finally science efforts are underway to confront one of the least talked about vulnerabilities that face our global interests.

True, we may not have a catastrophic encounter with an incoming space mass for another million years or we could face obliteration tomorrow at half past four.

"The Random Event" is why we buy insurance. There is no price tag on stopping NEOs coming our way.

These bold, badly needed, initiatives carried out in the absence of atmospheres or magnetospheres once again demand all our ingenuity to find ways to protect our heroic men and women when and if they are called upon to participate and address the serious requisites of finding ways to save the planet.

Chapter Eight

UNDERSTANDING SPACE BIO-HAZARDS

(This chapter is written specifically for those with science interest wishing to understand the radiobiology of space in greater detail. It is all GRAY and you should feel free to proceed to Chapter 9 without concern.)

(34)(27)(also see Appendix A.)

·······

GRAY:

By now the phrase Death Rays should be more than a literary device to get ones attention and this book has spent a good deal of effort describing their presence and origins so that it should be quite clear that this is a valid expression even though we have not yet detailed their absolute potential for compromising health.

There is ample historical evidence for the death of living cells due to radiation and that significant tissue and organ structures can lose their function either at the event of exposure or over a period of time leading ultimately to whole body extinction. It might therefore be convenient if all we need to know is the critical value necessary to avoid a specific degree of damage. Unfortunately the cellular impact of IR does not work this way.

The problem is that the overwhelming amount of data comes from research and epidemiological studies both human and animal all carried out on mother earth using low LET (energy transferred per unit of track in KeV per micrometer) X-Ray and Gamma Radiation.

Except from some recent data to be presented there has been very little genuine high LET studies comparable to space radiation and therefore much of our serious concerns involves the questionably valid extrapolation of one to the other.

But, it represents what we have in place up to now and we must proceed with what seems reasonable because many lives will be at stake.

There is a need then to be sure there is understanding what the term "death-ray" means in respect to the radiobiology of short-term and long-term survival, and the need to apply what we can to present and future space flight.

There is also a need to spell out what we do not know, what kind of transitions are being made from animal data to human data, where we must wait and see versus what we could know right now that is evidence based; and to put it all in perspective since the hazard exposure of all living things to radiation is clearly not an all or none affair.

We know there are some threshold differences as to how a person or persons will react to a given dose intensity (energy) and density (flux) and as pointed out that there are individual capacities to resist and recover. For example from animal data we believe astronauts in their fifties will have a better tolerance than ones in their twenties; that children and pregnant women will be particularly vulnerable in space flight or settlements. If these caveats are correct how will they effect space planning and colonization?

Dodging the Death Rays

The idea that spontaneous repair is a free trip to safety is a serious miscalculation. It is certainly true that much repair after exposure can be totally restorative however one of the confounding problems both in predicting the future is our implicit uncertainty that today's recovery is tomorrow's path to health.

Ionizing radiation may randomly destroy any cellular molecule within its track producing a failure of function. But our interest does not stop here. Our focus is also very much on the destruction of somatic (body) genes or sexual (ova or sperm) genes and the manner in which the injury (the mutation) is expressed either at once or over time that brings about such disorders as leukemia or a failed pregnancy or a deformed fetus.

Without plowing through the sordid history of radiation miscalculations, accidents, and outright disregard to a radioactive environment we all know that there is a good side to the story and that modern medical radiotherapy technology can focus its beam to destroy offending malignant tissue. The American Society for Radiation Oncology has abundant data to show that with early stage diagnosis radiotherapy can selectively result in absolute cure.

On the other hand studies on the horrific atomic bomb casualties at Hiroshima and Nagasaki (conservatively placed at ~250,000) or those caught up in the residues of a faulty nuclear plant collapse tell an indisputable story that ionizing radiation can cause serious malignancies and bring human life to an abrupt halt (viz: the extensive Hiroshima/Nagasaki research data of the Japan/United States RERF group, and the recent follow-up studies of Chernobyl victims, the sharp increase in thyroid cancer and late leukemia, the 31 bone fide radiation related deaths, and the hundreds of workers under long term evaluation).

These are examples of IR in broad strokes but for a more fundamental understanding of the subject we must take a moment to narrow down.

Given that the human body is composed of cells (many trillion of them) we note that they differ in size and shape and depending on how they have been programmed by special directing DNA produced proteins determines where they will locate (i.e. a liver cell versus bone cell etc.) and how they are to function.

However, one of the many fascinating facts of life is that all the nucleated non-sexual cells in our bodies have the same internal structure, the same number of chromosomes and organelles and mechanism for reproducing themselves.

The rate at which they reproduce and the response they make to their various gene based signals for what functions they will possess is what distinguishes them from one another.

All of our cells participate in a very specific life cycle. Certain cells of certain tissues will be doing this while you are reading this sentence. Others may undergo the identical process but only after significant passage of time, weeks, months etc.

The cycle may begin soon after the previous cell division with all the chromosomal nuclear material in a resting state in a nondescript tar-like pool. Suddenly this pool begins to take on some shapes and size along with accompanying growth of all the nuclear and non-nuclear structures that make cells function. This is called G1, and if the cell had not opted to move out of the cycle to a more lasting resting stage it will progress to the S (synthesis) stage in which there is burst of biochemical activity leading to preparations for making exact copies of itself and sorting out the precise number that will make up the 46 chromosomes found in every cell of every tissue of every organ in our body. (23 matching pairs, half from mom and half from dad, the ova and testes being the only exception for they will carry half that number).

This done the cell cycle proceeds to G2 in which the cell and 23 chromosomal pairs prepares to double themselves so that every facet of its complex anatomy is ready and the now doubled chromosomes will be set to migrate to the mid line.

Finally the last stage, mitosis, wherein a spindle forms on the opposite poles of the cell and having attached its silk-like fibers to the waiting midline chromosomes pull each of them in opposite directions while the cell is dividing creating exact copies of one another.

I have spelled this all out because it is this amazing and wonderful orchestration that is wounded by ionizing radiation.

Each pair of our forty six chromosomes have distinct recognizable features and contain similar but not exact matching genes.

The fundamental material out of which all chromosomes with their genes are constituted is the information sequences provided by the four vital base units that make up our DNA. This DNA is a twisted ladder of sugar and phosphate holding these bases in delicate balance and when a highly energetic alpha particle, proton, or HZE, or energized neutron comes barreling in on the structure, the ladder can break down, the sugar and phosphate can split apart, the bases can go flying off. The critical transfer of cell Information and protein production (DNA to RNA to

Ribosomes) stops or is misread, oxygenation stops, nutrition stops, growth stops, replication stops- the cell dies at once, the cells dies after a delay, or if the injury is limited the cells may undergo faulty mutation or some degree of repair.

Sometimes the cell maintains function in spite of some structural initial injury, all of this depending on the nature and energy of the incoming ray; and depending on what stage in the cell's cycle the structure is being attacked, in this case no repair may even be needed.

Radiation destructiveness is most apparent in the doubling phase of reproduction so that any tissue that is constantly renewing itself such as sex cells, bone marrow, skin, hair and the complex lining of the gastrointestinal track will succumb to radiation long before such quieter reproductive tissue such as muscle, bone and nerve cells.

The special problem with nerve cell injury is unlike all other cells in the body. Adult nerve cells do not replicate. And although stem cells help in early life we are born with just so many nerve cells and when they die they are not replaced.

All of this implicit IR genetic damage should not minimize the fact that there is a concomitant amount of non-genomic injury going on that adds to the burden of cell survival and potentiates the genomic injury.

Briefly stated IR takes the water that makes up 60 percent of the mass of every cell and subjects it to chemical alterations that produce oxidizing events (appropriately called oxidative stress) that damage vital macromolecules necessary for cell survival.

Specifically regard for a moment how important the non-nuclear structure mitochondria is to each cell's survival. Its citric acid cycle and respiratory chain produces the key substance ATP which is the currency of our total body energy. If any of its multiple choreographed steps is compromised all cell life can come crashing to a halt so that there are all kinds of chemical defenses to repair the slightest blemish. But subjected to a high enough dose of IR all those defenses crumble.

Note that different disorders have different scenarios as to dose and follow up.

For example radiation sickness (to be discussed) has what specialist call "deterministic injury" meaning that it occurs at a given absorbed dose and becomes increasingly more severe as the dose increases.

On the other hand cancer and genetic injury are "stochastic" meaning the effects of IR are chance events with the probability of the effect increasing with the dose but the severity will be independent of the dose. The description then is in terms of risk, leaving an estimate of the probability that a given dose will result in the disease.

With this as background the most important take home messages in all of this is to recognize that much of the fundamentals of space radiobiology are distinctly different from that of radioactive radiobiology and the radiobiology of X-ray and Gamma Ray exposure we are familiar with here on planet earth! It is a different breed and it is more deadly.

Space radiobiology is unique and is dominated by the high energy charged particles from galactic sources, and to a lesser extent from the Sun and those sequestered in the Van Allen belt. As noted, It is high LET radiation (energy transferred per unit of track in keV per micrometer) in contrast to low LET XRAY and Gamma radiation.

Much of space radiobiology is new to us and troubling in that we don't yet have full knowledge of its implications. We are just beginning to be able to partially simulate some of these highly charged high mass particles in the newly designed and constructed super-accelerators (The Loma Linda University Proton Synchrotron and The Brookhaven National Laboratory) accompanied by a broad commitment of animal and cell research.

The essential difference between the rays of conventional radiobiology and that of massive charged particles imploding on tissues is how they deposit their energies. The former is diffuse and runs down exponentially as it penetrates tissue whereas the latter has a linear track of varying thickness exciting neighboring atoms into secondary ions to form all the way on its structured trip to its final grand loss of energy into the "Brad Peak". It is much more efficient. It is much more destructive. It is much less forgiving.

Highly charged particles at low dose or fluence have been observed to induce a "bystander effect" in which the "hit" cell and many of its "non-hit" nearby neighbors react to the radiation inducing a wealth of activities, such as up-regulation of certain cell factors, cell cycle arrest, apoptosis (programed cell death), mutations etc.

These are some features of the new radiobiology of space and the problem is that we do not have adequate data yet to know its true impact on human health.

In the words of expert Dr. Gregory Nelson, "... it changes our traditional concepts of dose."

Dodging the Death Rays

This much we do know. Space ionizing radiations have the capacity to kill cells. These are the "death" rays.

These principles understood we are now ready to look at the specific major threats of space ionizing radiation

Chapter Nine
WHY WE DODGE

What exactly is there about living moving reproductive men and women that make us targets for these noxious agents? The answer lies in the distribution of WATER, PROTEIN and DNA in our bodies.

We are made up of approximately 65% water and when ionizing radiation smashes through water it creates charged molecules called free radicals that are highly reactive and damage the protein based cellular structures (the mitochondria, ribosomes and vacuoles) and nuclear enclosed DNA that are vital to life.

Ionizing radiation can strike DNA directly, breaking the critical bonds. One break can often be repaired but two breaks and a whole chunk of important genes can be lost and with it the loss of function.

Remember that DNA is the vital information bank that make up our genes and the genetic code. When it is disturbed, copying and translation at the nuclear assembly line for producing the stuff of life is impaired or completely destroyed and these events are quite enough to induce a host of DNA mutations in the wake of significant energetic and dense radiation.

Why then isn't every ionizing radiation event met with devastating results? There are two answers with many unknowns.

We have 23 pairs of genes (up and down all 46 chromosomes) harboring critical coding DNA in every somatic cell (non sex cell) in our body but this DNA makes up only a small 2 per cent of the total cell DNA. This is a mysterious and big biological surprise. We now know that most DNA in a cell is overwhelmingly non-coding and therefore takes most of the IR hit. In the past we wrote this DNA off as "junk DNA" however in recent years we have come to realize that there is hidden functionality in some of this residua. How these functions are compromised by IR is not completely understood but we can assume therein lies some protection or potential trouble.

The second answer of course is the remarkable capacity for the fractured DNA to be able to repair itself, a healing process that at times appears to render serious destructive ionizing events inconsequential.

As for the important non-DNA structures such as the cytoskeleton, the membranes, the organelles, and the busy trafficking cytoplasm that undergo injury, they are all vital machinery of the cell that lack the reparative skills of DNA. They can break down and their story is all downhill.

Armed with these facts we can now examine the effects of IR and the possible induction of disease; disease induced by direct hits to living tissue with prompt or delayed expression and disease that may be handed down to the next generation by virtue of the insult to the sex cells.

Let us take a look first at the second of the two main concerns about IR, namely the perpetuation of disease through its hereditary effects. What does the evidence tell us, not the fruit fly and mouse data, but the human data?

Interestingly we have yet to see a dramatic onslaught of heritable disorders arising out of the devastation of all of the ground based nuclear bombing, the tests and accidents (24) (51).

Why is this? The specialists tell us that we need more time and follow-up and we will slowly begin to see the rise of hereditary disorders in these select populations. Certainly we have not seen any concrete evidence yet of hereditary mutations emanating from those in space flight as of this writing. We obviously need more sperm and ova studies as well as examination of their progenitor cells.

The answer is quite different when we address the second concern, the question of human illness acquired in outer space flight by the prime generation of astronauts who have been our main subjects.

As noted previously the numbers of humans exposed to IR out in deep space have been small and the time periods relatively short but we have had an extensive experience with man in earth, and near-earth orbits, and we have important animal and laboratory research that will be hard to ignore (27).

There is one glaring fact about most of the research we shall examine and that is until recently with the advent of the new powerful accelerators (i.e. Brookhaven National Laboratory) we have been unable to simulate the high energy HZE radiation that has demanded our attention so that many of the health deficits that we look at are engendered by radiation doses orders of magnitude below what astronauts will encounter on prolonged exposure in deep space.

Although space IR concerns cover an increasing menu of pathological entities we will have a brief look at the most disconcerting namely CATARACTS, NERVOUS

SYSTEM DISEASE, RADIATION POISONING, PREMATURE AGING, CARDIOVASCULAR RISKS and a special look at the big one- CANCER. Finally we will have a look at what may prove to be its most arresting acute concern and still poorly documented category- the potentiating effects of IR on the human stressed IMMUNE SYSTEM.

First and foremost since the results are clearly established by Rastegar et al.(41) we can say unequivocally there is an increased incidence and early appearance of CATARACTS developing in astronauts who have undergone relatively low doses of space radiation. The confirming studies have been conducted by F.A. Cucinotta et al. reporting from the NASA Johnson Space Center with a 30 year follow-up on astronauts who have undergone space travel of various kinds (10).

Astronauts are so well protected from any Ultraviolet intrusion that this not a factor as it is at earth base. (They are so well protected that they require mandatory supplements of Vitamin D in their diets.)

Cataracts of course are at most visually annoying, generally quite correctable and do not represent a life threat, but they are a hint that something more invidious lies behind the findings manifesting in the eye.

From the first days of space flight both at low earth orbit and out beyond the magnetosphere astronauts have complained of disturbing recurrent unpredictable flashes in their visual fields. These mostly white dot, dashes, or clouds, can at times occur as often as three minute intervals, are improved but not eliminated within the shelter of the ISS but not the traveling spaceship, and correlate best with high energy Cosmic Rays penetrating the vitreous humor of the eye.

These Cosmic Rays, primary and secondary, appear to be hitting on the visual receptors of the retina or impacting the optic nerve itself and appear to be innocuous, but they call attention to some worrisome studies relating to IR and the CENTRAL NERVOUS SYSTEM (8)

....

There are both animal and human studies that validate the effects of low dose ionizing radiation compromising cognitive function, both as an acute and delayed effect (19).

This has also been observed in numerous high dose mouse and rat studies, and after radiotherapy to patients with brain tumors, generating problems of orientation both visual and spatial, often with associated memory loss (58)(3).

Our classic understanding has been that the central nervous system is resistant to ionizing radiation requiring high dose exposure to implement neural death. Recent research however has found that the hippocampus, one of the primary structures in the brain involved in higher brain functions, contains radiation-sensitive cell fractions (48).

In one of the most provocative recent studies by Cherry et al. (6) on transgenic mice subjected to high energy HZE type exposure at the Brookhaven NASA Space Radiation Laboratory two very important findings were made. First, all exposed mice versus the control non- irradiated mice after a six months delay developed flagrant cognitive impairment. Second, and even more attention-getting was the appearance in exposed mice of accelerated histo-chemical and microscopic Ab plaque changes characteristic of the findings in early Alzheimer's disease.

The design of the study was to try and replicate high energy Cosmic Ray exposure in space flight. Although this work needs confirmation it raises questions for how long must we wait before IR impacting on the human nervous system will become apparent and what form will it take.

Although most IR space research on the nervous system has concentrated on the brain there are numerous ground based studies from therapeutic radiology reporting on marked impairments of the spinal cord and peripheral neuropathies. There is little reason to doubt that all neural tissues would be targets for HZE radiation (26).

...

Every Astronaut out for a space- walk knows the potential nightmare of the ACUTE RADIATION SYNDROME (4). This unwelcome specter of whole body exposure follows them as soon as they emerge from the relative shelter of the spaceship.

A major CME or high HZE flux coming in at an intensity around 0.5-1 Sievert could produce within minutes the telltale nausea and vomiting that tells the exposed astronaut he has been caught. This is not in itself a fatal exposure but raises the problem of serious lung aspiration of projective acidic vomitus within the confines of a space mask that cannot be removed.

The acute radiation syndrome is deterministic. Exposures of higher intensities present both immediate and delayed onsets. As the intensity moves up we have the following possibilities:

First, the "bone-marrow syndrome" peaking around 60 days with a 3-5 Sv exposure resulting in-

1. Loss of lymphocyte cells that are critical for the immune system and the consequent onset of uncontrolled infection.
2. Loss of clotting factors and blood platelets resulting in an extensive hemorrhagic menace involving skin, brain, liver and kidneys.
3. Rampant fever and dehydration.
4. Death usually from sepsis (whole body infection).

Or, the "gastrointestinal syndrome" with intractable nausea, vomiting, diarrhea, bleeding, electrolyte displacement and collapse occurring within one to two weeks; death a result of shock and sepsis.

Finally the "brain syndrome" with complete loss of central control cognition and coordination evolving into coma, seizures, convulsions and shock, 100% fatal.

At ground level in a specialized hospital setting there are many remarkable therapeutic agents and supportive measures that can be instituted to modulate the dreaded Radiation Syndrome but up in outer space, beyond the point of return, the astronaut is out of luck.

The diagnosis of chronic radiation syndrome is worth taking note of because it is quite distinct from the acute form. It can occur at much lower doses of IR and is composed of a constellation of health effects with an onset of many months to years after exposure. An example would be the highly secret Russian Kyshtym reactor explosion in the '50s that resulted 20 years later in an exposed population outbreak of extensive skin atrophy, fibrosis and scarring (36)(23).

....

The culminating possibility of PREMATURE AGING, DELAYED DEGENERATIVE and CARDIOVASCULAR DISEASE brought on with month to year- long space accumulated exposure to ionizing radiation has had surprisingly little public attention, but there is accumulating data to support the contention. Most of the conflicting evidence has related to low energy radiation which is not our focus. What is clear, however, is that cosmic radiation and energetic particles from a solar CME have energies far superior to what we are recording from clinical and experimental data.

In a major review with 165 peer reviewed references Richard Richarson PhD. (42) summarized the entire subject with the following conclusions:

"The case is well documented for radiation-induced aging at high doses. It appears to be associated with free-radical damage to cell structures, double stranded breaks in nuclear DNA, increased programed cell death (apoptosis), and inflammation."

Which brings us to the remarkable contribution to cellular performance called telomeres. These are repetitive DNA sequences (TTAGGGxn) that cap the ends of all chromosomes protecting their gene integrity and orderly behavior. Over time these special end DNA sequences become shorter and shorter as a function of normal aging and ultimately lead to natural cell death (apoptosis).

A vital enzyme called telomerase restores these sequences and prevents the cell from aging but ultimately over time loses out. Paradoxically, there is some evidence in vitro that HZE radiation can in some circumstances up-regulate telomerase and prevent telomere destruction, (see Cancer beyond in this chapter). On the other hand progressive telomere shortening predisposes to increased sensitivity to IR, a provocative story that if correct could help explain the aging problem and early death from ionizing radiation (18).

All in all it is ironic that Albert Einstein predicted that after years of an extensive space trip an astronaut would return to find he was still young and his peers had aged. Old Dr. Einstein was unaware of the potential hidden deviltry of ionizing radiation in extended travel in space.

....

As for Cardiovascular Disease, when we look at data from the Japanese atomic bomb survivors we find an excess risk of delayed cardiovascular disease to doses far below that which will be found in outer space.

J.E.Baker et al. (1) from The Medical College of Wisconsin in a comprehensive recent paper, "Radiation as a Risk Factor for Cardiovascular Disease" summarized the peer reviewed evidence (105 research papers). The section on space radiation upon extrapolations made from extensive animal data in which serious delayed coronary disease was clearly demonstrated they make the following comment:

"The risk of developing coronary heart disease from radiation exposure during and after deep space exploration from Galactic Cosmic and Solar Particle

Events needs to be defined before attempting exploratory missions to the Lunar and Martian surfaces."

Certainly we have not the slightest clue so far that the moon mission, shuttle and International Space Station projects have revealed any statistically significant CV disorders. Death from coronary artery disease was an important casualty on the list we have but again the numbers are small and are overwhelmed by the accident data. The problem with going from IR exposure to the pathological findings of coronary artery disease is the "delay factor" of not weeks or months but years, the confounding intervening environmental factors and our failure to able to explain the biochemical mechanisms involved in the delay. Nevertheless we have a plethora of animal data that says IR can generate important inflammatory vascular precursors and that this is a real phenomenon (1).

Is the evidence strong enough to delay the space program ten or twenty years? Maybe not, but it introduces one more clinical entity that will require careful post flight follow-up and preventive treatment if necessary.

....

By far, most of our concentration as to why we need to dodge the death rays has been centered on the possibility of CANCER induction by space ionizing radiation and particularly high energy Cosmic Radiation (8).

We start with the recognition that all cancer is a stochastic effect of radiation- meaning that malignant onset has only a PROBABILITY of occurring as a opposed to deterministic effects which always happen after reaching a certain threshold (as discussed above with Acute Radiation Sickness). So the entire discussion of cancer causation and expression has this statistical basis.

We also start with the recognition that as of this date we have no data on cancer incidence on human exposure to high energetic Galactic Cosmic Rays sustained in space flight over time.

Furthermore characteristically there will be a delay in the expression of the disease, months in the case of hematologic disorders such as acute leukemia, to years in the case of other solid organ involvements.

In addition there will be selectivity in expression, children and women more susceptible, younger men more likely to succumb than older men, certain individuals more resistant to radiation effects than others.

Finally, there is simply still a void in the final answer to the interdiction and eradication of malignancy.

Here is what we know. Low LET X-Rays and Gamma Radiation can cause an extraordinary collection of malignancies and that this radiation is orders of magnitude below the much more damaging high LET Cosmic HZE radiation with the intuitive probability that the later will result in a greater incidence of human cancer.

Once again we have to turn to animal research and the wealth of information of IR human induced malignancy where the cancer risk is extensive for doses above 50 to 2000 mSv.

The studies of human exposure at Hiroshima and Nagasaki, at the Chernobyl reactor collapse, nuclear reactor workers and radiotherapy patients from five countries establish increased morbidity and mortality cancer risks at 12 tissue sites with bone marrow, lung, breast, stomach, bladder and liver heading up the list. All of this is ground based data (20)

Yes, the cancer/ space discussion is all about uncertainty. It is about calculated risk estimates complicated by the time of exposure, the age and sex of the crew, their relative health, family history, and the different IR environments they have encountered. It is full of imponderables and requires long follow-ups and large statistical samples to validate the conclusions.

In large epidemiological meta-analysis studies on airline pilots and flight personnel exposed in the attenuated upper atmosphere there is an increased mortality rate from malignant melanoma and an increase in female breast cancer (50).

The Nordic Study of Pukkala et al. (40) reviewing a cohort of flight personnel, 10,051 males and 160 females for a 20 year follow-up, found significant increases in malignant melanoma, squamous cell and basal cell skin cancer.

Not all studies have confirmed this.

In a search for the published and unpublished cohort studies of flight personnel from 1986-98 Ballard et al. (2) found an increased relative risk for melanoma, brain, prostate and breast cancer. They did not find support for previously described increases in acute myeloid leukemia. Importantly, they also noted that confounding issues such as found in non-occupational risk factors could contribute to the increases. This is the Achilles heel to all such clinical cancer epidemiology.

Because the question is always raised let us take a look at the actual astronaut mortality data as of September, 2014. Does it tell us anything?

Note at the onset. Of the 51 U.S. career astronauts who have died (Mercury through Skylab, Shuttle and Space Station) 27 have been killed in accidents.

Only twenty four men have ventured out beyond near earth orbit for a relatively brief period of days. These are the Apollo Moon-mission astronauts. They have undergone the type but not the duration of exposure an astronaut will encounter in the many months of flight and settlement time proposed. Twelve of the astronauts walked on the surface of the Moon and twelve remained in transport vehicles. The longest was Apollo 17, the final mission lasting 8 days.

Of those who walked on the Moon eight remain alive in their late seventies and early eighties with four deaths; one motorcycle accident, two heart disease, and one leukemia (at ages 69, 61, 82 and 74).

Of the twelve Apollo astronauts who remained in transport vehicles nine remain alive also in their late seventies and early eighties with three deaths; one from pancreatitis, one heart disease and one with a strange nasopharyngeal metastatic cancer at ages 61, 56 and 51 respectively.

This represents two cancers within the group of twenty four with most men living to flourishing old age.

Similar remarkably upbeat data comes from examination of shuttle participants who unlike the

Apollo group were exposed to modified IR for extensive periods at low Earth orbit.

There have been 135 NASA shuttle missions with 306 men and 49 women predominantly U.S. professional astronauts. Aside from the tragic losses from the Challenger and Columbia explosions there have been six death from cancer (two from malignant melanoma, two gastrointestinal, one breast, and one metastatic undifferentiated malignancy); there were three cardiovascular deaths.

And last but not least of the 140 U.S. participants on the International Space Shuttle we find only two deaths- one a plane crash accident and one malignant brain tumor.

What inferences can be drawn from this select population? The statistician will tell us that nothing useful can be drawn from this small sample, that we will need carefully collected data in large numbers over several decades.

However we note in this small number the high ratio of cancer to cardiovascular disease (2x the normal reported in the general population) and the possibly significant change in cancer type (2 brain, 2 bone marrow, 2 gastrointestinal, one breast, one rapid undifferentiated malignancy and two of the rare metastatic malignant melanomas).

It is clear that it will take time before meaningful cancer data will emerge from the astronauts who will venture into outer space or the larger group who will continue to circle in low earth orbit.

Meaningful cancer data will have to include what predisposing radiation exposure the astronaut has undergone previous to space flight. As noted, careful recording of the family history, what chemical exposures, and what markers of susceptibility (which are already understudy for some malignancies) need to be in place. All are important factors.

Today, cancer research has brought us out of the quagmire of yesterday's ignorance and we have enough clarity as to the mechanism of malignancy to have reasonable concerns about the death rays. Before leaving this central motif have a look at our present understanding.

Cancer is all about unrestrained cell division, a loss of the normal controls particularly for stem cells. Healthy stem cells pervade all tissues and are the "young pups" that will replace all the dying cells that constitute the life cycle of every tissue. All our tissues have variable growth cycles and repair processes. The more rapid the turnover the more vulnerable for malignancy. Included in every cell cycle scenario is the orderly programmed death of cells (apoptosis), and when indicated subsequent renewal.

The telomers referred to in the previous section on aging could play a significant role in space carcinogenesis. The enzyme telomerase that preserves and restores the aging cell can be rescued and undergo up-regulation by HZE radiation and thereby immortalize a cell- the hallmark of malignancy.

In another scenario, well identified agents (a virus, a chemical, a death ray) can target critical genes which when mutated turn on runaway reproduction. We call these oncogenes.

In addition these provocative agents can attack normal suppressor genes that ordinarily control the cell cycle and induce an orderly death of cells. Mutated suppressor genes (such as the P-53 gene) again produce unrestrained "immortal" cell cycles.

These agents can also affect genes that generate factors that take over the cellular environment so that cell growth becomes not only unrestrained locally but break down cell barriers, invade blood vessels so that the malignant cell spreads throughout the human body (metastases).

It is important to point out that the above processes frequently work in tandem since we believe that cancer cause is multifactorial often requiring multiple "hits" over extended periods of time.

IR is random in relation to cellular injury. It does not seek out a receptor to trigger a negative outcome. Depending on your view from the bridge it might be said that the cells' elaborate architecture and encapsulated chromosomes with their orderly placed nuclear genes and important extra DNA structures like mitrochondria (a critical cell energy pump) simply get in the way of the high energy incoming particles. If the incoming energy is strong enough and dense enough we have a statistical setting for damage and provoked showers of newly formed ions capable of creating highly reactive free radicals that cause additional cell injury. The normal repair mechanism are themselves altered and therefore fail in their mission.

With IR induced cancer the severity of the disease is independent of the dose but the probability of its occurrence increases with the effective radiation dose. IR does not affect the rapidity of the growth, the symptom profile of the tumor or even the prognosis. So here we are again having to settle for IR cancer induction in probability terms. We have to estimate. What, we ask, is the risk that a certain degree of exposure will bring about a neoplastic transformation? Are risk estimates binding? What are the confidence limits?

NASA has set a risk estimate for a fatal cancer from space exploration at 3 per cent. The Zeitlin Southwest Instrument Group that carried on recent live RAD detector data during the trip to the surface of Mars has allowed for some meaningful calculations. Thus a 360 day round trip using state of the art travel time yields an exposure of 662.4 mSv. (independent of the six months to a year planned on the Martian surface or with extra-vehicular activities). This alone increases the chance of getting a fatal cancer to about 4 per cent in a male and 5 per cent in a female. NASA quite properly will not accept this.

Thus, the IR cancer story is a glaring relentless challenge to all investigators that will require a much richer collection of laboratory and epidemiological data.

....

Finally we must ask a most compelling question for which at present writing we have only fragmentary human data and a wealth of animal data.

How can the ionizing radiation of extended deep space flight and surface settlement exploration directly or synergistically effect the HUMAN IMMUNE SYSTEM?

To best understand the impact of ionizing radiation on this system let us take a brief look at what is involved here.

In order for the human body to maintain its integrity so that all of its many functions work effectively over time it must be able to recognize and defend "self" from "non self". This is no easy job living in a world where there is a constant torrent of relentless enemies that can enter and destroy us.

We have the bio-torrent of viruses, bacteria, molds, yeasts, parasites and the like; the plant-torrent of spores, weeds, pollen etc.; and a host of non-living elements and molecules that are capable of causing serious disabilities.

Nature has provided us with three levels of defense against these threats to our immunity.

The first and oldest is the anatomical structure that turns out to be one of the largest organs in the human body- our skin and mucus membranes. An unbroken skin and smooth lining to our gastro intestinal, respiratory and genitourinary tract is a gift. There is a tendency to take this amazing capability for granted. In good health foreign material can enter us only if allowed.

The second level is what we have labeled The Innate Immune System. Penetrate the skin or mucous membranes with a foreign body and within minutes there develops a constellation of swelling, pain, redness and increased heat. It is what we recognize as inflammation. This is a specific response of a family of "white cells" each with special functions that intend to identify the intruder, recruit more cells to the location, call out specific relevant chemicals and destroy and remove the noxious material before it can invade the blood stream and cause septic shock, or pneumonia, or liver abscess or other bad outcomes.

Finally, there is the third level of immunity, a late comer in our evolution, highly sophisticated and complex: The Adaptive Immune System.

This is the system that calls out cells that can recognize the most minute offending agents, develop armies of specific antibodies to inactivate a vast menu of offenders, and develop cells that have memory so we are prepared for any such repeat. This is the system that make vaccines work, and in failure wipes out populations via epidemics, a system that signals with warning allergies and autoimmune disorders. This is the system that demands the proper match for blood transfusions and any attempt of an organ transplant. It never sleeps, it never rests and it is unforgiving.

Space flight has the capability of compromising every level of the immune system that we have just described. The scientific literature is replete with animal data in which the immune cells under study are significantly decreased in prolonged flight and the longer the trip the more aberrant are the data. The critical immune cell used for studies is the lymphocyte.

Research by K.George et al. (Mutat Res 2013) studied lymphocyte chromosome damage in five ISS astronauts before and after their first and second long duration space flights. All astronauts showed significant, but recoverable, increased exchange damage and translocations. Studies using the high charge high mass radiation seen in outer space result in more profound types of damage.

In August 2014 a NASA Bulletin called attention to studies in which the immune system is "dazed and confused during space flight" (12).

Certain immune cells were losing their function while others were hyper-functioning. NASA scientist Dr. Brian Crucian (7) notes that the system is being altered by many factors in the spaceflight environment including Radiation, Stress, Sleep Disorders, Microgravity etc. The longer the trip the worse it gets. Certainly that fits with space radiation.

Recent immune studies with astronauts who undergone extensive ISS flight have revealed that multiple latent viruses can reactivate during space shuttle missions

In October 2014 S.K. Meta et al. (28) who have published warnings for the past decade about viral behavior in astronauts in flight published an intensive study of 17 astronauts on space shuttle missions, found 14 subjects activated and shed in their saliva and urine the Cytomegalic Virus (capable of causing a wide range of infections), the Eptstein-Barr Virus (the cause of mononucleosis and lymphoma), Varicella-Zoster virus (the cause of shingles).

What could this mean in terms of frank illness, what form or forms will it take, how long in travel time and residence in space before we might see overt expressions of infectious disease? These are the haunting unanswered questions.

What is important to us in regard to our review is what specific role will ionizing radiation have in disrupting the role of human immunity in extended space flight and settlement since IR is clearly one of the known offenders?

The worst scenario seems repeatedly to point to a combined causation of microgravity, physical and emotional stress, nutritional factors and ionizing radiation (45).

NASA's challenge will be to mitigate each of these.

....

How all this concatenation of negative events explored in this chapter will end up, and with what degree of impact and impairment to the traveling and exploring astronaut it will have, we repeat, is what we do not know for lack of more definitive data, but we hear the hoof-beats, they are loud and clear and whether or not they are "horses" or "zebras" we are going to have to pay careful attention.

What is also clear is how great the need to structure that data as close as possible to meet scientific standards.

For example, if you are going to send men and women into space for extended periods of time the ideal would be a matched control candidate with the same profile and training who does not go out on a mission, in which there will be careful short term and long term health follow-up for each cohort and meticulous correlation with all radiation events and detection. This would take time and numbers and each group would need to be chosen randomly (the exception being the rare opportunity using identical twins such as now in process with the Kelly twins).

A major weakness in the present follow-up data on astronauts is the acquisition of valid death information and pathology reports from surgical procedures during life after discharge.

One solution would be that no astronaut goes into deep space without signing an agreement for an autopsy at the time of his or her death, whenever that may be, with all data and tissues available to NASA's medical department and a similar agreement to cover all health issues and surgical material during lifetime follow-up. Every astronaut needs to know that they are in this follow-up arena for life.

This is a feasible scenario. It is a must if we are going to make scientific sense out of what we clearly don't know.

All we absolutely do know now is that deep space exploration has inherent unavoidable risks and that the intellectual ferment surrounding the specific risks of ionizing radiation and human ventures in space requires cool heads and honest appraisal.

Chapter Ten
DEFENSE

NASA has a stellar safety record. From its very beginning it has had a serious interest in the health and welfare of its people. It has supported a vast array of mechanical, chemical and biological initiatives to bring meaningful mitigation of the space risks.

In addition there has been numerous inputs from the medical aviation community so that many of our space physiology problems have been well addressed if not totally solved.

There remains nevertheless the gnawing problems of ionizing radiation in space that will not go away.

The fact that we have been focused on the IR problem in deep space should in no way imply that we should not be concerned about what happens to the human body exposed to low levels and high levels of low energy and high energy IR in suborbital space acutely and over time? We have left this subject for others to address (9).

We have been centered on what happens to the human body exposed to low and high levels of low and high energy IR acutely and over time in deep space and we need to examine what we can do about it.

The first and most definitive mitigation is of course total abstention; that is, to substitute robotic mechanics wherever possible for human participation, to which we hear groans of disappointment because it removes the daring-do swashbuckling promise of man conquering the space wilderness. And one may ask, is Robotics a serious alternative?

Many distinguished scientists, engineers, politicians and surprisingly astronauts have lined up to support this alternative assuring us that it could provide for increased options and opportunities. Robots can work in any radiation environment safely and without fatigue. They can rove over vast areas, stay put, get into and out of difficult loci, they can undergo endless updating and repair. They have many shapes and sizes.

We have evolved over millennia. Space Robots have evolved over only a few decades with expanded uses and capacities developing almost monthly.

The main objection to Robotics, as the arguments bounce back and forth, is that human hands-on exploration is more adaptive and nimble, able to make critical decisions on the spot.

Compared to human creativity robotic activity is by nature stilted and lacks originality. There is always the problem of the failure to respond to the unpredictable and the inevitable time lag of communication in deep space.

As of now a robotic program is already functioning on Mars albeit only a start with a litany of possible expansions for the near future. But plans and planners for human exploration still keep knocking on the door.

The solution for future missions appears to be the development of a conjoint cooperative arrangement. "Robonauts" with arms and legs designed by NASA engineers will operate under the direction of live astronauts. It's the stuff of early space fiction. These mechanical wonders will do the dangerous work in space. They will save lives and burdensome labor. According to a recent NASA news release many have already had their tryout at the ISS (32).

Therefore we have a rational plan for their future. Robonauts may work all day outside in space unaffected and undeterred by showers of high energy Cosmic Rays. We will not have to watch the clock or fret over shelters to protect them.

The second defense which has been in operation since the first space ship left ground is shielding. Traditional spaceship shielding is caught between the dilemmas of mass versus efficiency. A three inch wall of lead in space to neutralize the effects of Cosmic Radiation is not an option. At present the aluminum walled spaceship can obstruct most solar radiation but that is only half the battle. It will not obtund Cosmic Radiation and particularly HZEs and of course we must not forget the astronaut on extravehicular assignments out in space.

For this we turn to the materials engineers who have struggled to provide safe supplies, water, food, mechanicals, laboratory equipment and oxygen for people all housed in an aluminum structure with a weight that allows for a safe send-off and that has substantial sheltering capabilities from a great deal of the ionic radiation of low orbit space (i.e the International Space Station) but cannot do the job in deep space.

New materials in plastic chemistry, taking advantage of the protective effect of hydrogen rich configurations, are believed to have greater shielding power and are in hand and in development but again do not yet have the quality to completely stop HZE Cosmic Radiation.

By 2005 NASA scientist Nasser Barghouty and colleague Raj Kaul had developed a polyethylene based material RXF1 that has superior structural properties and shielding properties than aluminum, and raised the question of constructing plastic covering or walls for a spaceship (33) but there has been little public comment in follow-up.

In June 2013 NASA released observations from instruments aboard the Lunar Reconnaissance Orbiter that plastic shielding indeed "reduces" the radiation dose from Galactic Cosmic Rays. How this material will play out over long spaceship trips and as protection for extra vehicular activities and settlements is still a wait and see affair.

Will we solve the shielding problem? Will the astro-chemists come through for us with a major breakthroughs? Can we save our explorers by having them carry around more effective protection? Hiding in caves, or keeping shelter away from the Sun on the frigid opposite side of the Moon or the canyons of Mars certainly hardly speaks for developing a comfortable community in space.

Still another avenue for defense has been developing some type of magnetic shield that would envelope the moving spaceship.

In 2005 Eugene N. Parker (37), a world expert on interplanetary magnetic fields reviewed the options in

Scientific American. The central idea in all proposals has been to nullify incoming positive charges.

He points out that a suitable magnetic field could evolve from a large centrally located electromagnet which would effectively deflect incoming charged particles. The required weight however stops the project at the door and to interdict HZEs the need would require $10 < 9^{th}$ electron volts and no one knows what the bio-effects would be on astronauts living in a magnetic field of 20 teslas.

Proposals that involve efforts to create extended field using plasmas proved not practical and the concept of "coating" the spaceship with a bath of positive charges fell flat when the ubiquitous negative charges in space were understood and the excessive power requirements factored in.

All in all magnetic field proposals were in limbo.

However, In 2008 through 2012 British scientists at the Oxford Appelton Laborastory make claim to a "mini magnetosphere" wrap-around for a space vehicle that provided almost complete protection from charged solar events, all emanating from a desk sized generator that can be switched on and off as the occasion demands (13).

We await more details and the opportunity of production to clarify its handling of CMEs and HZEs.

It is difficult to visualize space exploration with astronauts cowering under cumbersome shielding as the design for the future. The new observations are most encouraging.

The final option for defense from the medical standpoint is most exciting and challenging-namely progress in the development of pharmacologic radio-protective agents that either- (1) increase the resistance to the noxious products of ionizing radiation injury by neutralizing them or-(2) unravel the particular complex pathways of DNA repair due to radiation and find agents that preserve and enhance the healing pathways.

Recall that all the negative effects of ionizing radiation on DNA are under a constant predisposition to be repaired. A major method of healing is by natural intercellular chemical protectors ie. glutathione and polyamines. They are referred to as free radical scavengers and can inhibit the nasty destructive action of the reactive oxygen species induced by radiation. These are chemicals that can be synthesized and administered in pharmacological doses. How often, how much, and how safely we need to learn (54).

In 2008 J. Langell et al.(25) identified 15 such agents with significant radiation dose reduction from 1.1-2.4. Of these the significant stand out drug was an aminothiol chemical named Amifostine. It is now FDA approved. It has some significant toxicities that detract from its virtues. It requires intravenous administration. It best be given one half hour before exposure and for best results in high doses. This obviously leaves much to be desired.

In 2013 Anthony D. Kang and colleagues (22) submitted data on a protective agent (0N01210.Na) that prevented radiation that produced damage to mice and bone marrow cells exposed to lethal doses of ionizing radiation. It showed great promise as a novel agent for protection and treatment.

As of December 2014 Russian scientists have submitted encouraging data on the drug B-190, in which they demonstrated radioprotection and good tolerance, with the intent of combining it with other radio-protective agents (53).

Take note that many toxic drugs that have never been approved have also served as precursors for the development of many new important effective ones.

The second pharmacologic strategy is much more demanding and the research though intensive has not yet arrived at the summit.

It involves identifying the many protein agents in the different pathways that nature uses to reattach the DNA breaks induced by different offenders. When these promoting agents do this efficiently and effectively they can normalize the cell. When they do this incorrectly and inefficiently the results are a chaos scientist call "genomic instability". We end up with genetic mutations that are out of reach of cell survival and ultimately produce organ and systemic failure.

The good news is that some of the main pathways and their agents required to initiate chromosomal repair have been isolated.

A major problem is that the pharmaceutical industry is reluctant to invest in research in this area. It will be very demanding, expensive and have a relatively small population utilizing whatever drug emerges out of the trials. Specialized drugs are the markers of what all pharmaceutical companies shy away from, requiring run-away costs, prolonged recruitment and a host of potential litigation.

NASAs research allocations are spread thin. It is not the last occasion where economics will prevail over science.

In addition to the above defense items there are advocates for extending the biochemical events of "hormesis".

This has been a subject of considerable debate, contending that exposure to small amounts of radiation in selected individuals allow them to build up a resistance to ionizing radiation over time, analogous to an immunological technique. Find the secret of "hormesis" extract it and pass it on in pill form to anyone going into space. Great idea, yes, but unfortunately to date there are no firm evidence based studies to prove this phenomenon exists and at present has been rejected by much of the radiobiology community.

There is however the classic observation that some human beings have a better tolerance for radiation exposure than others. Just what this means at the biochemical level is not well understood but may represent resistance to breakdown rather than the increasing ability to repair.

If some identifying marker for these individuals can be forthcoming it could affect the selection process for flight and of course if some unique chemical factor could be identified pharmacologic research would have a new target.

Finally it is not unexpected that in the absence of real solutions for carefully researched drugs to fill this critical gap the inevitable non peer reviewed "outsider" will creep into the picture.

In 2012 The New York Daily News and the UK The Mail reported on a wonder "space drink" AS10 "developed by NASA and confirmed by research at the University of Utah". It repaired space radiation induced wrinkles and reduced UV dark blemishes in a four month trial group of mostly women, and close to 200 participants. It was the cosmetic answer to aging.

The drink was made from a blend of exotic fruit and other plant derivatives. It was available online. A 12 day supply would cost about $50.

Within days of the report NASA made a vigorous response that this was pure fabrication. NASA had nothing to do with the spurious "space drink" and it was definitely not a NASA product. A University of Utah spokeswoman similarly stated emphatically that no such research had been authorized by the University. That the claimed author of the research was a former "volunteer" faculty member who was no longer associated with the institution.

So much for pharmacologic miracles.

But the valid research effort is telling us something positive is in store. Once again we shall have to be patient and await real meaningful breakthroughs.

Chapter Eleven
FINAL THOUGHTS

If one stays out in the rain they will get wet. Staying out longer they will surely get wetter, longer still they will be soaked and could be hit by lightning.

Out there in deep space it is raining high speed radiation composed of protons of varying energies. They are unique and unlike any of the natural radiation we know here on earth. Unlike wet rain their shower never lets up but unlike wet rain it cannot be seen and it cannot be felt so there is a predisposition to under-rate it. That is a serious mistake.

High mass proton radiation damages human cells and we have several trillion cells in our bodies. The question that emerges from space radiation is how many human cells have to be destroyed by high mass radiation before the body will recognize this as illness or disease? Or how many brain cells, or bone marrow or heart cells need to be compromised before there is organ impairment, or worse, before all spontaneous repair processes have been overwhelmed?

In a comprehensive review by Space Medicine experts Chancellor, Scott and Sutton (5), Life, 2014,4(3) they note, regarding Galactic Cosmic Radiation, the following:

"During transit outside of Low Earth Orbit, every cell nucleus within an astronaut would be traversed, on average, by a hydrogen ion every few days and by heavier HZE nuclei (e.g. 0-16, Si-28, Fe-56) every few months. Therefore, in spite of their low flux, HZE ions constitute a deleterious biological threat and contribute a significant amount to the GCR dose that astronauts will incur outside of Low Earth Orbit."

Who are the individuals raising the flag of caution and labelling space radiation the number one risk to astronauts in deep space? I will give you a sampling while staying close to the NASA family.

Francis A. Cucinotta, formerly Chief Scientist at NASA's Space Radiation Program, now Professor of Health Physics at the University of Las Vegas has been an important steady contributor to the literature on this subject, warning about the risks for years.

Janice L. Huff, Ph.D. Deputy Element Scientist, NASA Space Radiation Program issues a recent comprehensive document entitled "Space Radiation and Risks to Human Health.

Jeffrey P.Sutton,M.D.,Ph.D. Director of the new National Space Biomedical Research Institute writes "with long duration missions the physical and psychological risks to astronauts are significant".

Note that Sutton is a coauthor with Jeffery Chancellor of the NSBRI and Graham Scott, of Baylor College of Medicine in a remarkable comprehensive paper with 83 peer reviewed references entitled "Space Radiation: The Number One Risk to Astronaut Health Beyond Low Earth Orbit."

The list goes on and on.

This is a sample of scientists who are telling us that we do not have all the answers but we know enough to establish clear and cautious warnings since there is reason to believe that many of our risk estimates are flawed and do not fully embrace the new space radiobiology (5).

We knew enough to warn about the dangers of smoking before we understood the molecular pathways of lung cancer.

Is this all scare talk? Yes, it certainly is, because the evidence is scary.

Given the facts of the space environment which has been spelled out in detail and given the imponderables that still exist, what would constitute at this time a sensible deep space program for the next decade?

The answer will not have a pleasant sound to the many persons whose agenda has been fixed and who are lined up and ready for blast off with a year or two travel time in space and months in resident habitats.

There are multitudes of both eager young and old professionals who tell us that they will take the risks whatever they may be but they do not want to miss the opportunity of making space history.

There are astrophysicists and engineers who have devoted their careers to making the Moon, Mars and the Asteroid trip in space the premiere scientific achievement of the century.

There is a cornucopia of dollars already woven into these considerations with public office holders and private investors regarding these space adventures an important economic opportunity and development and of course jobs.

Lastly, there is a host of sincere patriotic citizens out there who, whatever the costs may be, want to beat the Chinese and beat the Russians at the gate before they make their move.

But, the facts of the risks to human health and life are in bold type and we are morally bound by an old but indelibly wise medical Hippocratic admonition that says "Primum Non Nocera", (first, do no harm).

We need evidence that human beings in extended time and travel (months to years) in outer space, exposed as they will be to the unique ionizing radiation that is out there, will not acquire acutely or in a delayed manner the pathological incursions of degenerative disabilities, premature aging, coronary artery disease, cognition and neuronal failure any or all of which the research suggests could await them. Not the least of our needs will be factual data that put the risks of the dark panel of malignancies and immunological disasters to rest.

We are all too cognizant of the fact that a vast effort is going on hoping to mitigate these risks; that there will be more breakthroughs in shielding material, planning changes to accommodate the solar cycle, improvements in radio-resistant therapy, and earnest fixes of every description. But as of this decade as we prepare to send the astronaut(s) into deep space we need a major rethink.

We do not at present have enough detector information on deep space ionizing radiation to measure with confidence the overlapping risks. We do not know how to precisely predict the time or intensities of the provocative Solar Particle Events, although some recent evidence claims this may improve. We do not know the price the astronaut body will pay for cumulative exposure. We surely do not have control of the ever present shower of GCRs and their HZEs, or how to effectively manage them. The risk estimates for the most serious illnesses are in dispute.

Given the information we do have here is a reasonable scenario:

The waiting game can be painful but it pales before a commitment that confirms our worst fears. The slow deliberate search for an honest protective technology if it can be found will help generate the proper tools for future plans in outer space.

The argument for robotic exploration is unassailable. Talk with robotic engineers and they will tell you they are ready to go now. Every day new ideas and new improvements emerge from their research on robotic capabilities. We may lose a great deal of money heading their way but we will not lose lives.

The argument for going to the Asteroid and the Comet since they are the true enemies of our planet and it is worth risking lives to find a solution to their imposing threat of global mass extinction- this argument has much strength and deserves consideration if the actual requisites for the presence of human input holds up under scrutiny.

Dodging the Death Rays

The ultimate safety of planet Earth as formulated in asteroid exploration trivializes any human venture at this time to the planet Mars with its long trip and barren explorations which have a paucity of meaningful additions to our proximal welfare and therefore for now should be and can be left to robotics which will increasingly prove their merit.

A changed policy would start with the Moon based project only if it is clearly essential as a base for the more important asteroid venture and identification of Near Earth Objects.

Trips and stays on the Moon would need to be layered and made relatively short. They would not be the objective but the tool to master the incoming massive objects.

The Moon is comparatively close and has a better prospect for rescue for an injured or sick astronaut, but prolonged extravehicular activities would still lean on robotics as the primary input and move carefully with human participation so we can buy time and ultimately have humans in space working safely with and directing the mechanicals.

Little room has been given to proposals for permanent stations or communities. They have the clear disadvantage of placing our species in impossible environments that we have spent eons of evolution to overcome. They should be dropped for now.

Little has also been said about lower energy proton exposure for those astronauts not in deep space such as the ISS or those continuing their activities inside low earth orbit. This book would leave that for others who have raised serious questions for those taking repeat trips and stays (9).

We need to give some serious thought to the legal and ethical implications of promoting international commercial outer space travel before the question of the medical risks of ionizing radiation are settled.

NASA and the international government-run programs have historically chosen their candidates for space flight with great care and concern for their preflight, flight and post-flight development and welfare. NASA has a record of constant inquiry into every positive or negative element of space flight. Its research programs are being constantly concentrated and upgraded. As noted above NASA's own health experts have repeatedly made unambiguous warnings about all of what has been presented. But NASA has a mission and it remains to be seen whether the mission transcends the facts of this book. It also remains to be seen if the private enterprise cadre is listening to these warnings since wrapped around these space dreams is a litany of potentially litigious problems.

So far much of the tenor of human commercial space flight advertisement is upbeat. With the dollars in place it is a concern that lobbying will commence to free up these promised voyages into space from restrictions, regulation and surveillance.

It is somewhat remarkable that in all the discussions on space exploration we have had little mention of an item called "the quality of life". Much has been said about the awesome beauty and vastness of the moonscape or the excitement of a walk or an extravehicular maneuver looking back at the tiny blue planet called Earth. Little has been noted about loneliness, stress and depression, real problems that have had to be confronted since the beginning of space flight.

The space physician hears a voice in the background that asks:

What does it mean to spend our efforts and treasure on a plan that places man in an unforgiving hostile environment, living in caves with elaborate requirements for food, breath and shelter, dressed in elaborate cocoons with a need to be on constant alert for an enemy that cannot be seen or felt, that is deceptively silent, constant and will come after you from any direction?

What kind of space society will it be to have our pregnant women hidden from the light of day and our children restrained from running freely out in the open spaces? What would it mean to live in a world of unremitting anxiety where for twenty four hours seven days a week you are on the alert?

The philosopher asks, "Is this a life?" The psychologist asks, "Can the functioning brain and the human spirit sustain itself in these barren forbidden settings over time?"

Finally, you the taxpayer will soon have a vote and you will need to take a stand. You will make your own decision, but hopefully you will take advantage of the facts that have been presented.

There is a need here for mature minds. We know we cannot change the laws of physics. If we are going to go into deep space for real we shall need to know precisely where and why we are going, what we are going to have to face, the opportunity costs involved and the particular requisite of exactly how we are going to dodge the death rays.

REFERENCES

1. **B**aker JE, Moukder J, Hopewell JW. Radiation as a risk factor for cardiovascular Disease. *Antiox Redox Signal* 2011 Oct 1 15(7) 1945-1950
2. **B**allard T, Lagorio S, DeAngelis G, et al. Cancer Incidence and Mortality Among Flight Personnel: a Meta-Analysis. *Aviat Space Environ Med* 2000 71: 216-224
3. **B**elka C, Budach W, Kortmann RD, et al. Radiation Induced CNS Toxicity – molecular and cellular mechanisms. *Br J Cancer* 2001; 85; 12
4. **C**enters For Disease Control and Prevention. Acute Radiation Syndrome: A FACT SHEET FOR PHYSICIANS *CDC* 2014, Oct 17

5. Chancellor JC, Scott GBI, Sutton JP. Space Radiation: The Number One Risk to Astronaut Health beyond Low Earth Orbit. *Life 2014, 4*(3), 491-510; doi:10 .3390/life4030491

6. Cherry JD, Liu B, Frost JL, et al. Galactic Cosmic Radiation Leads to Cognitive Impairment and Increased Ab Plaque Accumulation In a Mouse Model of Alzheimer's Disease. *PLOS ONE, 2012, Vol 7 issue 12 p1-9*

7. Crucian B, Simpson RJ, Mehta S, et al. Terrestrial stress analogs for spaceflight associated immune system dysregulation. *Brain Behav Immun* 2014 Jul,39:23-32

8. Cucinotta FA, Durante M, Cancer risk from exposure to galactic cosmic rays: implications for space exploration by human beings. *The Lancet Oncology 2006, Vol7 Issue 5 p 431-435*

9. Cucinotta FA. Space Radiation Risks for Astronauts on Multiple International Space Flights. *PLOS\one* 2014 April 23- DOI:10,1371/journal.pone.0096009

10. Cucinotta FK, Jones MJ, Iszard G, et al. Space Radiation and Cataracts in Astronauts. *Radiation Research 2001*, 156, 460-466

11. Dainiak N. Biology and clinical features of radiation injury in adults. *UpToDate, 2014.*

12. Dunn A, Abadie L. Study Reveals Immune System Dazed and Confused During Spaceflight. *NASA News* August 18 2014,

13. Eiscat. Shield for the Spaceship Enterprise, A Reality? *Appelton* Laboratories *magnetic shield for space radiation.* National Astronomy Meeting 2007

14. Gabrielsen, P. More from Crater on radiation Health Hazards *Lunar Networks* November 18, 2013

15. George K, Rhone J, Cucinotta FA. Cytogenetic damage in the blood lymphocytes of astronauts: effects of long-duration space missions. *Mutat Res* 2013 Aug 30, 756(1-2) 165-9

16. Golsden Spike Company announces Phase A lunar landing mission *NASA spaceflight.com 12/6 2012*

17. Gopalswarmy N, etal. Screaming CMEs Warn of Radiation Storms, Research quoted in *NASA Bulletin FEATURE 05, 29 2007*

18. Goytisolo FA, Samper E, Martin-Caballero J, et al. Short Telomeres Result in Organismal Hypersensitivity to Ionizing Radiation. *J Exp Med* 2000 December 4, Vol 192 Number 11, 1625-1636G

19. Greene-Schloesser, Robbins M. Radiation-induced cognitive impairment-from bench to

bedside. *Neuro-Oncology* 2012 Vol 14, Issue Suppl 4, iv37-44

20. Greene T, Latowsky G, Silver K. Cancer and Workers Exposed to Ionizing Radiation 2003 May, *Center for Environmental Health Stidies.*

21. International Commission on Radiological Protection, *3rd International Symposium on the System of Radiological Protection,* 2015

22. Kang AD, Cosenza SC, Bonagura M, et al. 0N01210.Na (Ex-Rad) Mitigates Radiation Damage Through Activation of the AKT Pathway. *PLOS /ONE* march 7, 2013 DOI 10.1371/journal pone.0058365

23. Kaplan J, Toxicological Profiles for ionizing radiation *US DEPARTMENT OF HEALTH AND HUMAN SERVICES* 1999

24. Kodaira M, Satoh C, Hyama K, Toyama K. Lack of effects of atomic bomb radiation on genetic instability of tandem repetitive elements inn human germ cells. *Am J Hum Genet* 1995 Dec;57(6) 1275-128

25. Langell J, Jennings R, Clark J, Ward JB Jr. Pharmacological agents for the prevention and treatment of toxic radiation exposure in spaceflight. *Aviat Space Environ Med* 2008 Jul, 79(7): 651-60

26. Liang BC. Radiation-Associated Neurotoxicity. *Hospital Physician* 1999, April; 54-58
27. Maalouf M, Durante M, Farley N. Biological effect of Space Radiation on Human Cells: history, advances and outcomes. *J Radiat Res* 2011; 52(2); 126-46 Review
28. Mehta SK, Laudenslager ML, Stowe RP, et al. Multiple latent viruses reactivate in astronauts during Space Shuttle missions. *Brain Behav Immun* 2014 Oct: 41: 210-7
29. Modisette JL, et al. Radiation Plan for the Apollo Lunar Mission. *AIAA paper69-19 Rocket and Space Technology* 1969
30. NASA feature November 8, 2005. *Sickening Solar Flares.* Comments by F. Cucinnata
31. NASA Newsletter June 27, 2014
32. NASA International Space Station. Robonaut-2. *NASA News* 2014
33. NASA Headline News- A "designer material" derived from plastic...August 25, 2005
34. Nelson GA. Fundamental Space Radiobiology Gravitational *and Space Biology Bulletin 2003*, 16(2) 29-36
35. Olinto AV. Ultrahigh Energy Cosmic Rays and Neutrinos, 2011, *Nuclear Physics B Proceeding Supplements,* Vol. 217, Issue 1, p 231-236

36. Olson M. International Mayak Action Day *Nuclear Information and Resource Service* 2010
37. Parker, En. Shielding Space Travelers. *Scientific American* march 2006 39-47
38. Portree D. Humans To Mars: Fifty Years of Mission Planning 1950-2000
39. Available as *NASA SP -2001-4521*
40. Pukkala E, Aspholm R, Auvinen A, et al. Cancer Incidence Among 10,211 Airline Pilots: A Nordic Study. *Aviat Space Environ Med* 2003 Jul 74 (7) 699-706
41. Rastegar Z, Eckart P, Mertz M. Radiation-induced cataracts in astronauts and cosmonauts. *Graef's Archive for Clinical and Experimental Ophthalmology 2002* Vol 240, issue 7, 543-547
42. Richardson RB, Ionizing radiation and aging: rejuvenating an old idea. *AGING* 2009, 1(11) 887-902
43. Russian Federal Space Agency Report – Cosmonauts to the moon. *Space.com 3/5/12* 2012
44. Smith DS, Scalo JM, Risk due to X-Ray Flares during Astronaut Extravehicular Activity. *Space Weather 2008*, Vol, XXXX DO1:10:1029
45. Sonnenfield G. Immune Responses in Space flight *Int J Sports Med* 1998 Jul 19; Supp 3:S 195-202

46. Strange Adario. World's Apace Agencies Plan Future Joint Manned Mars Mission, *Mashable* 194, December 2014
47. Straume T. Ionizing Radiation Hazards on the Moon. 2008 *NSLI Lunar Science Conference*
48. Suzuki K. Neurotoxicity of Radiation. *Brain Nerve* 2015 Jan 67(1) 63-71
49. The Mars 100: Mars One announces Round Three, *Amersfoort,* and February 2015
50. Tokumara O, Haruki K, Bacal K, et al. Incidence of Cancer Among Female Flight Attendants: A Meta- Analysis *J of Travel Med* 2006 17 May Vol 13 issue 3
51. UNSCEAR *2001 Report* Hereditary Effects of Radiation; Report to the general assembly by The United Nations Scientific Committere
52. U.S. Department of State, Office of the Spokesperson. International Space Exploration Forum Summary, January 10, 2014
53. Ushakov IB, Vasin MV. Radiation Protective agents in the radiation safety system for long-term exploration missions. *Human Physiology* 2014 Vol 40, issue 7, 606-703
54. Von Deutsch AW, Mitchell C D, Williams CE, et al. Polyamines protect against radiation-induced

oxidative stress. *Gravit Space Biol Bull* 2005 Jun; 18(2):109-110
55. Wired.com/2013/11/ ICE CUBE TELESCOPE FINDS HIGH-ENERGY NEUTRINOS. OPENS UP NEW ERA IN ASTRONOMY.
56. World Nuclear Association, Nuclear Radiation and Health Effects. Updated January 2015
57. World Nuclear Association, Radiation and Life, December 2012
58. York JM, Bievans NA, Peterlin MB, et al. The Biobehavior and neuroimmune impact of low dose ionizing radiation. *Brain Behav immune* 2012 FEB 26(2) 218-227
59. Zeitlin C, Hassier DM, Cucinotta FA, et al. Measurements of Energetic Particle Radiation in Transit to Mars on the Mars Science Laboratory. *Science* Vol.340 no. 6136 pp1080-1084, 31 May 2013

BIBLIOGRAPHY

Primary peer reviewed articles will be found on Pub Med in abstract or full text

NASA Archives

NASA Bulletins

NASA Letters

CANCER Letters

Science

Scientific American

Astronomy magazine

Sky and Telescope

Astronomy Today Chaisson/McMillan Prentice Hall 4th Ed.

Solar Astronomy H- Beck et al William Bell,Inc

Principles and Practice of Nuclear Medicine Early & Sondee Mosby 2nd Ed

A Clinician's Guide to Nuclear Medicine Taylor et al. The Society of NM

Decisions and Evidence in Medical Practice Gross Mosby

High-Yield Cell and Molecular Biology Dudek Lippincott,Williams & Wilkins

The Timetables of SCIENCE A.Hellemans & B. Bunch Simon & Schuster

APPENDIX A. SPACE RADIOBIOLOGY

The radiobiology with which we are most familiar is of course derived from our local natural radioactive background and artificial medical diagnostic and therapeutic applications. These are for the most part rays from radioactive decay from unstable isotopes seeking to decay down to stable elements or from focused beams from Orthovoltage units.

Alpha Rays (helium ions) unless ingested and inhaled have a limited destructiveness due to their lack of penetration.

Beta radiation (emanating from the nucleus of radioactive elements) travel short distances in tissue and are most damaging when ingested or injected by a radioisotope that homes to a specific target, viz. radioiodine-131 to the thyroid cell.

X-Ray and Gamma Rays emanating from radioactive decay are capable of destroying cells and tissue by virtue of their high energy photon state interacting with the outer orbital electrons of atoms and generating ionizing components. This photoelectric effect is the dominant form of matter interaction for both X-Ray and Gamma Rays. Gamma radiation has a higher degree of activity and depth and is emitted from the nuclei of unstable atoms whereas X-Rays emanate from the electron environment. More recent data supports the recognition of much overlap between the wave frequencies of these two entities.

When emitted from astronomical sources such as gamma ray bursts, pulsars and quasars, the energies involved shift to immensely higher order states.

Radiobiology is concerned with how much ionization takes place in and around the cells of exposed tissue and at what depth from the incident ray. It is also concerned with how much energy is lost to the medium from a given incoming ray. This of course translates into the disturbed chemistry generated and consequent cell-tissue-organ damage.

A conventional dosimetric term used in the book has been the Linear Energy Transfer (LET) which reveals how much ENERGY an ionizing particle transfers to the mass per unit of its penetration and depends on radiation type and the nature of the challenged mass.

For example beta radiation has low LET profile as does X-ray and Gamma radiation, massless photons without charge whose LET values are based on the energy lost by the secondary electrons they generate.

Alpha rays on the other hand have a high LET profile, are very destructive but as noted barely penetrate the skin.

Another concept is the Relative Biologic Effectiveness (RBE) used to compare the effectiveness of different types of radiation on different tissues or cells (X-rays being the traditional standard).

The importance of separating out this earth based understanding is that Space Radiobiology is separate and distinct.

Here the environment is seeded with solar protons traveling in the solar wind and those trapped in the Van Allen Belts, with massive outputs of charged particles from solar flares and coronal mass emissions.

And superimposed on this ionic spread are the much more energetic and fluent high LET Cosmic Rays characterized by fully ionized atomic nuclei of all stable elements dominated by hydrogen, helium, oxygen and iron but capable of fully ionizing elements higher up the periodic table, producing an assortment of highly (H energized (E) high proton value (Z) ions, the unstoppable unavoidable highly destructive HZEs.

The three properties of space charged particles are:

1. They deposit their energies along well defined linear tracts rather than the diffuse fields characteristic of X and Gamma Rays.
2. They have a defined range in matter.
3. They produce secondary particles via nuclear interaction.

This structured energy pattern allows for levels of action at the molecular level, cellular level and tissue level that can induce damage that evades repair systems.

Most importantly from the standpoint of the health considerations of space flight are the expert comments of Gregory Nelson Ph.D. of the Loma Linda University Radiation Program:

> "Traditional concepts of dose and its associated normalization parameter, RBE, break down under experimental scrutiny, and probabilistic models of RISK based on the number of particle traversals per cell may be more appropriate."

High LET radiation produces unique patterns of:

DNA damage
Gene expression
Mobilization of repair proteins
Activation of cytokines
Remodeling of the cellular micro-environment

There are other characteristics of space radiation that complicate RISK assessment such as the "bystander "effect referred to in the chapter 8 and a possible multigenerational delay in the expression of radiation-induced genetic damage which is not dose dependent.

Francis Cucinotta, Ph.D., the prolific space radiobiolgy expert, in a recent monograph questions many previous risks evaluations and notes,"… qualitative differences in the biologic effects of Galactic Cosmic Rays compared to terrestrial radiation may significantly increase these estimates and will require new knowledge to evaluate."

This knowledge will come from new experimental data derived from the high accelerator technology now available.

We are thus still in transition as to the relative probabilities of illness facing our men and woman in deep space and much of what we have previously assumed will need to be moved up in risk assessment.

Dr. Gregory Nelson has an excellent paper in Gravitational and Space Biology Bulletin 16(2)2003 as well as Dr.s Francis Cucinotta et al. article in PLOS one. October 16,2013.

APPENDIX B. TIMELINE FOR COSMIC RAYS

1785 Charles Coulomb's well shielded electroscope spontaneously discharged. What was this penetrating depolarizing mystery?

1895 William Roentgen discovers X-Rays

1896 Henri Bacquerel accidentally discovers radioactivity and it fosters the idea that all air ionization was due to the ground based minerals.

1909 Theodore Wulf notes higher levels of radiation at higher levels. His work is questioned.

1912 Victor Hess- makes extensive balloon study measurements. He concludes, "..radiation of very great penetrating power enters the atmosphere from above". He rules out the Sun. He is awarded The Nobel Prize (1936).

1914 Werner Kohlhorster finds increased ionization as far out as 9 kilometers.

1920 Robert Millikan coins the term "Cosmic Rays". He believed the primary ionizing rays he observed from elevations down to deep water detectors and were caused by hydrogen fusion products and Compton scattering from outer space and that they were Gamma Rays.

1927 J. Clay corrects Millikan. He finds that Primary Cosmic Rays are deflected by geomagnetic fields are therefore charged particles. Notes that cosmic flux also varied with latitude.

1929 Bothe and Kohlhorster confirm Clay's work. Charged Cosmic Rays can penetrate gold.

Bothe tries to localize their direction in vain.

1929 Dmitri Skobeltzen finds some Cosmic Rays occur in groups called showers

1930 Bruno Rossi predicts the "east-west effect" deflection of Cosmic Rays by earth's magnetosphere with east having greater intensity and being positive. His work is quickly confirmed.

1931 Bruno Rossi proves Cosmic Rays can penetrate 1 meter of solid lead.

1932 The Millikan-Compton debate, Gamma Rays versus charged particles. Compton wins out.

1934 Walter Baade and Fritz Zwicky state that supernova alone could account for the high speed and energy of Cosmic Radiation

1945 Balloon research with photographic emulsions indicate space accelerators are generating heavy molecules from Cosmic Ray collisions.

1948 Physicists note HZE's in Cosmic Radiation raise radio-biologic concerns.

1952 Tobias predicts visual flashes in space flight from Cosmic Radiation.

1952 Donald Glaser observes Cosmic Ray tracts in his bubble-chamber.

1954 Chase describes graying of hair balloon-borne black mice.

1954 Walther Bothe receives Nobel Prize for his work on Cosmic Rays.

1955 Eugster demonstrates cell death in vitro from single hit of heavy ions.

1961 Brustard - cell death from heavy ions on maize embryos.

1963 Brain injury studies on balloon-borne mice.

1965 F.Reines & J. Sellshop detect Cosmic Ray Neutrinos in a South African mine trial a breakthrough for neutrino research.

1970 Haymaker- brain studies of Cosmic Rays on balloon borne monkeys show acute changes

2012 Voyager 1 reaches interstellar space. There is a drop in the flux of solar radiation and increase in the flux of Cosmic Radiation.

2013 NASA and Italian observers find W44 and IC 443, two supernova remnants show definite Cosmic Ray signals.

2014 The U. of Wisconsin IceCube research project in Antartica reports on unique high energy neutrino findings from outer space. These have energies above 1 petaelectron volts with unknown bio effects.

APPENDIX C. COMMON HEALTH PROBLEMS IN SPACE FLIGHT THAT MIGHT POTENTIATE THE RADIATION HAZZARDS.

The idea that cumulative insults to human hemostasis can predispose to disease is as old as medical history. It makes sense to ask what happens when radiation effects are added to the multiple health problems intrinsic to space flight.

If indeed this is a valid juxtaposition or only limited to particulars such as the immune system it follows that putting a human in space over time is a built in clinical experiment that needs careful evaluation. Although we do not have the answer to the inquiry at present a brief review of some of these health concerns is important to keep in mind.

Astronauts in flight are exposed to a palate of inescapable environmental, physical and psychological stresses that can affect their health.

The following is a brief list of these factors:
1. ENVIRONMENTAL.
    Microgravity
    Oxygen requirements
    Water requirements
    Thermal adjustments
    Nutritional requirements
    Personal waste disposal
    Trash management
    Radiation
    Injury from a space foreign body

2. PHYSIOLOGICAL
    Acceleration, Vibratory, Acoustic
    Weight loss
    Fluid shifts to thorax and head
    Vestibular-Space Motion Sickness
    Loss of muscle mass- atrophy 5%/mo
    Loss of heart mass-return dilemma
    Weakened ligaments

Stress muscle tears
Uncontrolled muscle twitching
Low back, shoulder and pelvic pain
Increased height in hypogravity
Osteopenia-osteoporosis
Bone fracture
Kidney stones
Flatulence
Slow wound healing
Hematological changes
Microbiological changes
Endocrine changes

3. PSYCHOLOGICAL- short and long term
   Stress
   Anxiety and fear
   Psychosexual problems
   Sleep disorders
   Privacy concerns
   Depression
   Maladaptation
   Isolation

4. OCCUPATIONAL
   Trauma- inside crowding, flotation
   Trauma- outside-suit penetration with-

Loss of consciousness, 9-12 secs.
Death within 2 minutes
Hypoxia
Air Bubble Embolism
Skin frost
Barotrauma
Fatigue
Extra Vehicular Accidents
Multitasking

5. SOCIAL / CULTURAL
   Hygiene
   Male-Female relationships
   Ethnic and National Bias
   Religious Diversity
   Social Custom

APPENDIX D. QUANTUM COMMENTS

Quantum theory unlike Newtonian mechanics is full of uncertainties and probabilities and is the price we pay for moving from the ordered comfort of the old to the new physics. Electromagnetic radiations travel 2.99 x 10>8 meters/second as waves in a vacuum from source to target with perpendicular oscillations in an electric and magnetic field. These waves have built in relationships. Wavelength, the distance between peaks or troughs, is the reciprocal of the frequency, the number of cycles passing under a fixed point (one cycle= 1 is the Hertz).

These waves have amplitude, displacement up or down from the zero line with consequent changing energies.

The EM spectrum has a range on a log scale of 10^4 (Radio waves) to 10^19 (Gamma Rays), with visible light making up only 2.3% of the spectrum.

It is the behavior of these waves that highlights the bizarre nature of Quantum Mechanics since EM waves are also particles, discrete massless packets of energy (*quanta*) we call photons which harbor energy proportional to the wave frequency.

We will be centered on the extreme right of the spectrum in the loci of X-Rays (frequencies 0.3-30 million gigahertz) and Gamma-rays with higher frequencies and shorter wavelengths.

It is the "Hard X Rays" that command our attention and all Gamma radiation since they penetrate tissue with photon energies considerably higher than the chemical binding energies and can produce damage.

Energy contained in radiation is best expressed in terms of electron-volts (eVs), usually at the Kilo (thousand) or Mega (million) level. (one eV= 1.6×10^{-19} joules)(one joule = 6.24×10^{18} eVs)

X-Rays and Gamma Rays have energies running in 100's of eVs to 10's of MegaVs.

Here is a quick look atQuantitative Radiobiolgy.

1. Radiation science has been dogged by a series of changes in how to represent dose. The following data is from the Canadian Center for occupational Health and Safety:

Bacquerel (Bq): The SI unit of radioactivity
(1 event of radiation emission per sec.)
Curie(Ci): The non-SI unit of radioactivity
(1 Ci is a large amount of radioactivity)
1 Ci= $3.7 \times 10^{\wedge}10^{th}$ Bq
Gray(Gy): The SI unit for absorbed dose
(I Gy=amt. 1 joule /kg wt. of organ)
Rad®: The non-SI unit for absorbed dose
(1 Rad=tissue dose- XRay & Gamma Ray)
1 Rad= 0.01 Gy
Sievert(Sv): The SI unit for bio- abs
Rem(rm): The non Si unit for bio-abs
1 Sv=100 Rem ()

Most modern ionizing radiation literature will express the dose in Sieverts since it takes into account biological effects of different types of radiation. Human doses are usually expressed in millisieverts (mSv)

2. Energy of ionizing radiation is measured in electron volts. One electron volt is the amount of energy gained or lost by one electron moved across the potential difference of one volt. It is a very small unit and usually expressed as kilo-electron volts (KeV) or Mega-electron volts.

Finally, recognize that we are talking about matter at a level that is well above quantum field theory and the fundamental basics of the standard model of particle physics which has our utmost respect.

Now that the Higgs Boson is well established since the celebration in July 2012 physicists can talk with some confidence about the nature of matter.

Particles are really vibrations in fields that pervade all of space, fields that interact with one another and are constrained by underlying symmetries. It is the Higgs field that brings this all about, whose vibrations are seen as Higgs Bosons and endow all particles with mass. Electrons without mass would be flying off into space, quarks would not make up protons and neutrons, and neutrinos would be zero entities all whipping past each other at the speed of light. There would be no world or life as we know it.

The particles described in this book are the particles and EM waves of Nuclear Physics and represent a level of complexity well above the standard model. However, be assured, they will serve to carry the message of this book with fidelity.

www.ingramcontent.com/pod-product-compliance
Lightning Source LLC
Chambersburg PA
CBHW031054180526
45163CB00002BA/831